The Secret Life of
SALMON

This book is dedicated to Christopher L, Charlie C, Patrick B, Theo B, Henry F, Trond, Howard R, Neil B, Roddy, and other friends from the rivers.

And to my daughter, Pippa.

The Secret Life of
SALMON

Henry J. Giles

WHITE OWL
AN IMPRINT OF PEN & SWORD BOOKS LTD.
YORKSHIRE – PHILADELPHIA

First published in Great Britain in 2024 by
White Owl
An imprint of
Pen & Sword Books Ltd.
Yorkshire - Philadelphia

Copyright © Henry J. Giles, 2024

ISBN 978 1 39901 196 9

The right of Henry J. Giles to be identified as author of this work has been asserted by him in accordance with the Copyright, Designs and Patents Act 1988.

A CIP catalogue record for this book is available from the British Library.

All rights reserved. No part of this book may be reproduced or transmitted in any form or by any means, electronic or mechanical including photocopying, recording or by any information storage and retrieval system, without permission from the Publisher in writing.

Typeset in 11/14 pts Cormorant Infant
by SJmagic DESIGN SERVICES, India.
Printed and bound in India by Parksons Graphics Pvt. Ltd.

Pen & Sword Books Ltd. incorporates the imprints of Pen & Sword Books: After the Battle, Archaeology, Atlas, Aviation, Battleground, Discovery, Family History, History, Maritime, Military, Naval, Politics, Railways, Select, Transport, True Crime, Fiction, Frontline Books, Leo Cooper, Praetorian Press, Seaforth Publishing, Wharncliffe and White Owl.

For a complete list of Pen & Sword titles please contact

PEN & SWORD BOOKS LIMITED
George House, Units 12 & 13, Beevor Street, Off Pontefract Road,
Barnsley, South Yorkshire, S71 1HN, England
E-mail: enquiries@pen-and-sword.co.uk
Website: www.pen-and-sword.co.uk

or

PEN AND SWORD BOOKS
1950 Lawrence Rd, Havertown, PA 19083, USA
E-mail: uspen-and-sword@casematepublishers.com
Website: www.penandswordbooks.com

Contents

Foreword by Dr Paul Rouse, FRSA		6
Acknowledgements		9
Author's Note		12
Chapter One	Looking Through Water	14
Chapter Two	A Fish of Northern Parts	21
Chapter Three	From Home Rivers to a Secret Life at Sea	28
Chapter Four	Controversial Secrets: Fish Farming and Hatcheries	50
Chapter Five	Barometer of the Ecosystem	62
Chapter Six	Making a Difference	83
Chapter Seven	The World Fish Shows the Way	93
Chapter Eight	The Wild North, Nature and Self-Knowledge	104
Conclusion		111
Appendix One	Seasons of Change: A Year on a Salmon River	118
Appendix Two	Salmon Facts and Factoids: A Cheat's Guide to the Secret Life	123
Appendix Three	The Battle of the Smolt Trackers	131
Appendix Four	The Eternal Salmon	136
Bibliography		140
Index		141

Foreword

Salmon are exquisite, magnificent, diverse, and beautiful. The flash of sunlight off the flanks of a salmon as it leaps, twists, or turns quickens the heart and sparks the imagination. Yet, although we stop and marvel in wonder, we are complicit in driving the greatest threats to the species, in all its forms, that it has ever faced. You and me.

We have become the biggest driver of change in the Earth system. The influence of humankind on our planet's systems has become so pervasive and fundamental as to usher in a new geological era – the Anthropocene – reflecting the overwhelming biophysical effects we are having on the planet.

Humanity now manages three-quarters of all land, excluding the ice sheets and we exploit half of all fresh water on Earth. We have created the highest levels of greenhouse gasses in over a million years, and already heated the entire global climate by over a degree, with at least another half a degree locked in and inevitable.

We have created a hole in the atmospheric ozone layer; fix more nitrogen than all-natural processes; and we are causing many of the world's deltas to sink due to damming and mining.

Human engineering has become so large scale that individual projects are having a global effect. For example, the angular momentum of the water restrained by the Three Gorges Dam in China has shifted the Earth's axis by 0.8 of an inch and slowed the Earth's rotation. Humanity has even created a new form of rock, having made in the order of 500 billion tons of concrete since 1930, sufficient to place one kilogram of the material on every square metre of the Earth's surface – land and sea.

Salmon have been present on planet Earth for (in northern Europe) at least 100 million years. Over six epochs the species has flourished across multiple diverse environments, adapting, and evolving as new opportunities arose and others were closed off, diversifying to fill niches to expand population numbers and range.

That is a story of success played out over millennia. However, now that the Anthropocene is driving planetary scale changes faster than any time since the dinosaurs – as a direct result of our behaviour, our choices, habits, and practices – the Pacific and Atlantic salmon species are now facing extreme challenges, with greater demands upon species survival in the oceans, rivers, and lakes than they have ever experienced.

Atlantic salmon jumping the white water. (Adobe Stock)

The environment which supports the ecosystems which salmon are dependent upon is realigning. The drainage that feeds the rivers which salmon rely on to reproduce are altering due to our agricultural and other use choices – the landscapes that feed nutrients into rivers that give sustenance to salmon parr are fundamentally changing.

The temperature of waters in their range are changing, as is the salinity of the water. Sea level is rising rapidly, and deltas and estuaries are being flooded, denying the salmon critical conditions for their survival.

Meanwhile the wild fish are being exploited as a food source by humans in unsustainable numbers and our factory farming of salmon is creating a genetic crisis while concurrently poisoning the waters salmon migrate through and creating untenable parasitic loads. Furthermore, the number of the salmon's predators in the freshwater environment has exploded, with millions of fish-eating birds migrating to fresh water from the seas, driven by collapses in populations of their prey fish species due to our overfishing and the changes to the ecosystems that we are driving.

Our farming and industrial practices, and even our contraceptives, are polluting rivers, with catastrophic consequences – directly causing fish kills but also affecting genetic diversity and capacity to reproduce.

As we test the planetary boundaries of the living space for multiple species, including ourselves, we must decide whether we care sufficiently to respond and change direction, or whether we would rather stay on our current path, with the inevitable consequences of biodiversity collapse, catastrophic, irreversible climate change, and the threat of many other negative results.

We sit at a fulcrum where we must rapidly resolve humanity's intentions towards the Earth's future. We must in this immediate generation resolve to either respond

The southwest Miramichi, New Brunswick, is a pristine Atlantic salmon environment. (HG)

with a new thinking, and a new relationship with the only known planet in the universe to support life, or choose to blunder on to a dire outcome.

Survival will depend on our future approaches to our planetary stewardship. We must settle how good planetary governance is embedded into our daily life, our politics and economy in a new regime of active planetary management. Salmon are an important indicator species, one that shows us the implications of our choices.

It is a species that gives me hope, as does Henry Giles's book. If, through the chapters readers can become ever more entranced by the species and understand how it relies on our own daily choices and decisions, and if readers agree it is a sufficiently special creature to warrant protection, something of importance that goes way beyond the salmon might be achieved. Some readers may make changes in what they eat and the products and materials that they consume, helping the salmon survive the twenty-first century, but also benefiting the wider environment and the prosperity of future generations.

Dr Paul Rouse, FRSA
Lewis, Outer Hebrides, Scotland

Paul worked with the Carnegie Climate Governance Initiative as its Science Adviser, is a visiting fellow at the University of Southampton and recently joined the carbon dioxide removal hub at Imperial College, London.

Acknowledgements

Thank you, Paul, for your support and striking words in your foreword. It confirms my decision to focus *The Secret Life of Salmon* as a 'state of the salmon nation' call for change. Coming from your perspective as a climate change expert, not long after your stint at the Carnegie Climate Governance Initiative, your hard-hitting eco call to arms has power and resonance.

Researching and writing this book has marked a stepping up for me, in this same spirit. This is my second book. My first - *How To Catch More Salmon* (2018) - was a very different project, coming from my passion for salmon fishing with rod and line. However, while it concerned catching more salmon, it was as much about a salmon's secret processes. And it has been enlightening to note how natural it has felt to segue on to this more species-oriented and eco project. So while the process has felt 'natural' and logical, it has also firmly opened my intention to the green camp – certainly on climate change and its consequences.

Proposed after an invitation from the publishers in late 2019, just before the COVID-19 pandemic hit in March 2020, and written during its ongoing impact through 2021, a thought kept cropping up. The plight of the salmon and our own species are intertwined and recognition of the need to act to save the Atlantic, chum, sockeye, king, silver and pink salmon, has arrived not before time.

The Cains river in September, New Brunswick. (HG)

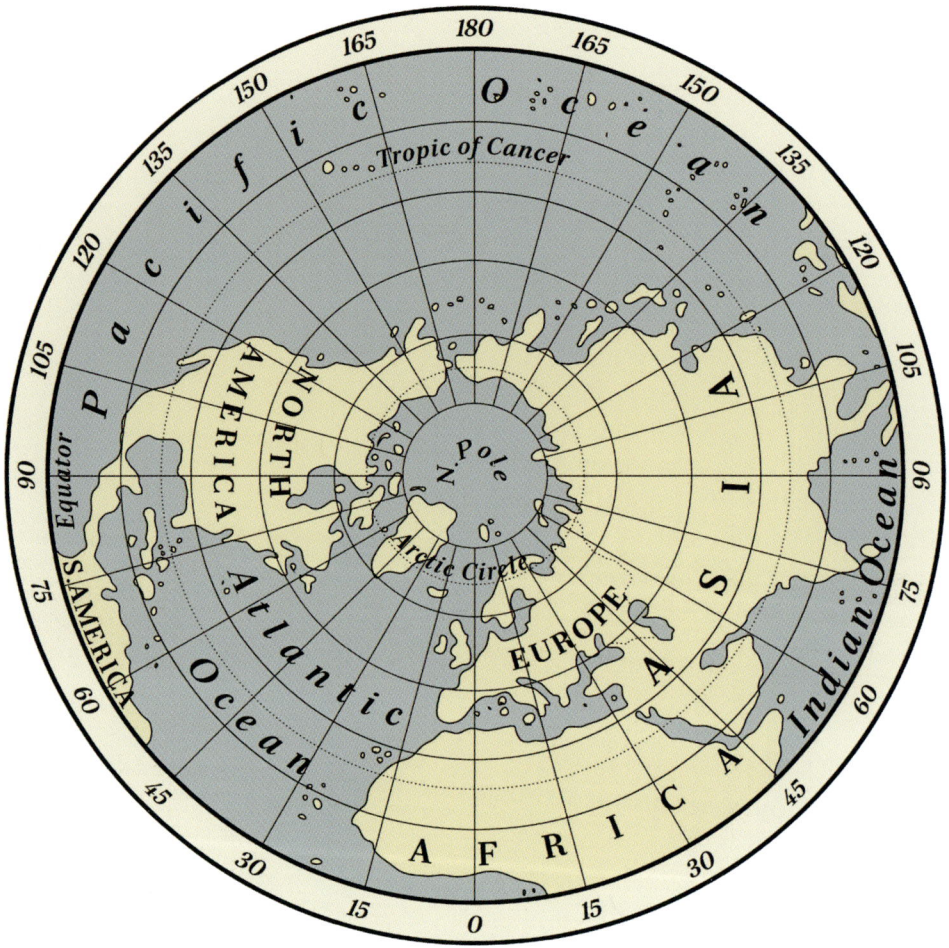

A stylised impression of the northern salmon waters. (Adobe Stock)

Thank you to Charles and Jonathan at White Owl Books/Pen and Sword Books, and to Carol, my editor, who re-worked, re-inspired and re-subbed large tracts of this book, offering help that was key in the editing stage. Thank you also to Janet, my production co-ordinator, and Charlotte, at White Owl. The whole team also supported me, and continue to support me on my first book, and the element of continuity has been special and welcome.

Just to acknowledge the inspiration of other writers on salmon, some of them in the bibliography here. Michael Wigan opened doors, with his radical and far-reaching account of the salmon story. So did Derek Mills with a relentless, and yet human-scaled, scientific backdrop. Peter Coates wrote one of those books that is so accessible and wide-ranging, it makes you think; what more is there to say? But I found a calling as the project developed and was astounded that the salmon's secrets revealed many more questions, and I hope answers to those questions, than I had dared dream of. Not least that the different species of salmon in so many ways have the answers, or at least an adaptive and maybe sustainable way forward. Yes, the fish have a plan! It felt like a story that needed to be told and I hope you agree.

A special mention to all those who get the importance of salmon and the marine ecosystems they depend on and are working on, or worked in their lifetimes, to try and save our planet while there is still time. A leading light was the late Orri Vigfússon, who was a key player in the process of pushing back in favour of Atlantic salmon, via his work buying out the drift-nets with the North Atlantic Salmon Fund.

I was lucky to meet Orri a few times. A member of my fishing club, I'll never forget his charismatic engagement (and childlike delight) showing me a new coffee table book on salmon – all in the name of NASF fund-raising – in the Long Bar of the RAC in Pall Mall a few years ago.

Thank you to Justin Maxwell Stuart, of Where Wise Men Fish, for his photographs of wild salmon country, and for his support on both books (Justin was one of the first to step breezily off Pall Mall into Farlows on the day of my first book launch, one of those morale-boosting moments you don't forget.)

Thank you to Chloe and Pippa, and my parents and family for your support, and putting up with living with a writer trying to get a book out, which is not easy.

Finally among the human species, thanks to Henry 'The Bear' Fosbrooke, who was with me at the start of my salmon journey on the River Nairn in 1980, forty years ago, for his wise words about the wise salmon. Sung as a ballad with a battered acoustic guitar, they made a hauntingly beautiful post-midsummer 2021 visit so special. Thanks also to Alec Laing at Logie Estate, Findhorn, for making that, and other visits, possible.

And thanks to the salmon, for direct contact through my fishing life, and bringing me close to the threatened but still wild places of the north. Bringing me peace, bringing us all peace, I hope, much needed in turbulent times. And that is why it is a fish worth fighting for.

Author's Note

There is a paradox in highlighting the secret world of salmon. The paradox comes because the fishing and wider media is full of the significance of salmon and its importance as the 'king of fish' – or the kings and queens of the river, to bring the metaphor into current usage.

In some ways a fish was never as well documented. TV documentaries, books, vast photographic libraries of stills and YouTube and other videos and wildlife and fishing movies are testament to the box-office draw of the salmon. Its dramatic lifestyle and specialised pull on our imaginations are there for all to see.

And yet there is a gap. A disconnect.

And this is the secret world that is the subject of this book. Or to be exact the secret world and a discussion or conversation or assessment of the broad places where the hidden world interacts with the documented world.

Why is it important to write of and document this? Because to build up a picture of the hidden parts of the salmon's world is to better understand the salmon. And with understanding comes our best chance of helping the very survival of this inspirational fish, such a key part of the freshwater and marine ecosystems. And through knowledge gained from a fellow highly evolved organism of our world, saving the planet. It really is as important and as existential as that.

The salmon has power on a cultural and emotional and spiritual level for humanity, too. The American writer Henry David Thoreau may have said 'most men fish all their lives without knowing it's not fish they're after' but what people do yearn for is interaction with wild nature. And here the salmon can help. This is where people can directly contact the wild experience of our hunting ancestors via the direct pull of a fish on a rod (that's really what they're going for, not the fish itself).

And armies, or at least militias, of eco conservationists across the northern Atlantic countries can get to grips with salmon, whether they fish or not, in a similar way. Perhaps cradling a CO_2-anaesthetised smolt in the palm of their hand as they measure it and attach a tag in the name of scientific research. Cradling it, holding it. In direct contact with this highly evolved life form with the potential to swim off to arctic waters then return as a two-foot-long bolt of silver.

So welcome to the secret life of salmon. It is secret, and it matters, so let's bring it out into the open where we possibly can, put the meat on the fire and feel the power, but also keep the important little bits of mystery. Because that can be a good thing too, amid the Atlantic and Pacific ocean miles we cover here.

Henry Giles
August 2024

CHAPTER ONE

Looking Through Water

The Dordogne is a mystic place, lost in the proverbial mists of time. It is wine country, as famous for its truffles as it is for its beauty. One of France's most legendary poets, the swordsman Cyrano de Bergerac, was born there. But long before Cyrano's seventeenth century, other men lived there in the caves that dotted the landscape around the rivers Vézere, Dronn and Isle. And in one of those caves, in the Gorge d'Enfer at Les Eyzies, is the first known human depiction of a salmon.

It is extraordinarily realistic, carved in bas relief into the chalk face. Above it are eight notches. Are these a reference to 'kills', a sort of prehistoric log book in which an early fisherman boasted of his catch? The man who carved the salmon was a hunter-gatherer. He lived with the natural world in a way we have forgotten, probably spearing salmon in shallow rivers for his own survival. What other relationship he had with the fish he had carved we cannot know, but we can understand at least that the salmon was a vital part of his existence.

From the Dordogne in France to the waters of the Atlantic and Pacific coasts, the salmon formed this important element in man's experience. As the hunter-gatherers of Les Eyzies developed a settled existence, grew crops and raised animals, the salmon continued its role as a source of food. The fish can be found in ice-fringed Arctic waters and swimming past the manicured lawns of Hampshire's River Test.

The relationship between *Homo Sapiens* and *Salmo Salar* is complex and of long standing. Among the littoral communities in northern waters, the respect that the Saami and the Inuit had for the fish they caught is reflected in the exquisite carvings of those First Nation hunters. It is still evident today, but has morphed into a far more scientific approach with built-in checks to avoid over-exploitation creating what we in the west call a sustainable harvest.

The First Nation peoples used ingenious hunting methods, using natural fibres to trap fish on the ebb tide. It is said that the word 'net' comes from nettles which made excellent weaving material – actually, it comes from the Latin *reticularius*. On Lewis in the Outer Hebrides, there are ancient stone structures in the Grimersta river which may be early examples of a salmon farm. On such waterways, it is still possible to fish with a nano-technology space age fly rod over structures built 6,000 years ago.

August Gaula Grilse. (HG)

Holding in the flow; biding their time. (Adobe Stock)

All over Canada and North America, especially in British Columbia, Alaska, Washington State and Oregon, communities settled at river-mouths and the debouchments of tributaries where they could catch salmon easily and exploit the fish. It provided food, oil and even clothing. There is ample evidence to show that these First Nation communities moved according to the salmon season, a unique type of transhumance, according to the different runs of fish on different rivers. This way of life had a rhythm of its own, a geographical gyration, meeting at 'the Table' as it was called in some cultures, but isolating in a 'secret dance' with the coho, sockeye, king, silver or pink salmon.

This complex relationship between man and fish is explained superbly by world expert Michael Wigan in *The Salmon* (2013). Because of its importance, the king salmon was a king-maker; at least a kingdom maker. The geographical pattern of river and the cycles of a salmon's life meant that the most usual form of transhumance – for example, the Plains Indians following the buffalo herds – were less traumatic and potentially destructive.

Centuries before man invented phrases like ecology and eco-systems, sustainability and even climate, our ancestors saw god-like beings in the natural things around them. Indigenous peoples looked to the salmon for spiritual guidance. Salmon bones were used for children's toys, fish skin for shoes and pouches. The first salmon ceremony was an integral part of the hunting year. It centred around the first catch of the first fish and was followed by the ritual of gutting, cooking and eating. Then the bones not used would be returned to the sea as an offering to a great, vast force not controllable and barely understood. It was, although our ancestors did not know it, part of the ecological cycle.

Tribes like the Yurok in the American north-west held back from harvesting the first fish runs in late spring until they were sure that the salmon were back in sufficient numbers to make fishing life-sustaining.

By the nineteenth century, after a number of false 'Renaissance' starts, science had established itself as the saviour, at least in theory, of the natural world. Frank Buckland was the godfather of fishing science. In *British Fishes* (1873) he wrote:

Doubtless, to each fish each river has got its own smell ... When the salmon is coming in from the sea he smells about until he scents the water of his own river. This guides him in the right direction, and he has only to follow up the scent ... to get up into freshwater ... So a salmon coming up the sea into the Bristol Channel would get a smell of water, meeting him; 'I am a Wye salmon,' he would say to himself. 'This is not Wye water, it's the wrong tap, it's the Usk. I must go a few miles further on,' and he heads upstream again.

It may sound a little patronizing today, but Buckland was trying to explain a complex phenomenon in terms which mere mortals could understand! Earlier civilizations had their own interpretation. The Northwest coast culture, including the Coast Salish and Haida peoples, had little agricultural land and relied heavily on the sea and its fish. One delicacy was 'salmon cheese', made by storing salmon in boxes until it achieved a runny texture. As Oliver La Farge wrote in *A Pictorial History of the American Indian* (1956) – 'Many white men have strong cheeses, but almost without exception we prefer cheeses that do not wave at us.'

The Northwest Coast communities believed that salmon were immortal. They believed the fish swam into rivers to feed mankind and bears, died and were reborn in the ocean. Out of this belief ritual became tradition, centring on the chiefs who owned the fishing grounds.

Such ritual and folklore was expressed in European communities too. The Irish folk-hero Finn MacCool, responsible for all kinds of shenanigans in the third century, recognized the 'salmon of knowledge' in a well. Instinctively, he grabbed it and its juices burned his thumb, which, equally instinctively, he put in his mouth. That, of course, gave him the salmon's wisdom and from that day, his power increased, enabling him to tell false kings from real ones.

The author returns a September grilse in New Brunswick. (HG)

In Norse mythology, from four centuries later, Loki had an argument with the blind god Höðr which led him to kill his brother Baldr (the beautiful). On the run, Loki changed himself into a salmon and leapt into a fast-flowing river. But the gods were still after him and they set a net to trap him. He leapt, as only salmon can, but he was caught by Thor who grabbed Loki's tail, which explains the tapering shape of the fish.

Science and awareness of ecology only just arrived in time. As with other wildlife, most noticeably the buffalo in the United States, excessive hunting was putting a huge strain on various species' survival. In his State of the Union address in 1908, President Theodore Roosevelt, who had something of a rapport with native animals, recognized the problem:

> The salmon fisheries of the Columbia River are now but a fraction of what they were twenty-five years ago, and what they would be now if the United States Government had taken complete charge of them by intervening between Oregon and Washington. During these twenty-five years the fishermen of each State have naturally tried to take all they could get, and the two legislatures have never been able to agree on joint action of any kind adequate in degree for

The secret life of a salmon running up shallow water close to the bank. (Paul Rouse)

the protection of the fisheries. At the moment the fishing on the Oregon side is practically closed, while there is no limit on the Washington side of any kind, and no one can tell what the courts will decide as to the very statutes under which this action and non-action result. Meanwhile very few salmon reach the spawning grounds, and probably four years hence the fisheries will amount to nothing; and this comes from a struggle between the associated, or gill-net, fishermen on the one hand, and the owners of the fishing wheels up the river.

Roosevelt was talking about a small part of the global reaches of the salmon. These fish once ran in every country with rivers opening into the Atlantic and the Baltic. Even Luxembourg, Switzerland and Bohemia (today's Czech Republic) had Atlantic salmon. Although now terminally entangled by dams, water abstraction and pollution, these fish runs came via the Rhine to be caught up in middle European countries without a sea border. There were also significant salmon runs up the Thames and Czechoslovakia's Vltara river. When Julius Caesar's legions invaded Gaul in 56BC, they found salmon leaping in the Garonne. In the Asturias region of eighteenth-century Spain, an enlightened study recorded 2,000 fish a day being caught in the rivers there. We have no way of knowing, of course, how accurate such a figure is.

Industrialization in all its forms has curtailed the salmon runs. In Britain, the Trent, the Ouse, the Ribble, the Clyde and the Taff were undermined by long-term and sustained levels of pollution. The Industrial Revolution saw mills powered by water springing up along such river banks and the output of coal and steel in the nineteenth century merely added to the problem. The same thing happened in the United States, where the rivers of Maine (once the 'salmon state') led to serious drops in numbers. From what had once been 17 per cent of the global Atlantic salmon run, the figure fell to less than 1 per cent, especially after the building of colossal dams along the Penobscot river. The same disaster has befallen New York State.

The salmon was always an emblem of wild places. The fish's life cycle is the original literary quest story – leave home, go on an adventure with the odds stacked against you, return by way of an emotional homecoming. It is a symbol of courage and strength. It is beyond eco. Beyond conservation. Beyond fishing. Beyond the outlying reaches of scientific discovery. We have – and have always had – respect for this wonderful fish with its perfectly drawn lines, smooth and proportioned like the reverse shear of an ocean-going yacht, it is part of folk memory and is a living example of the tantalizing thinness between the boundaries of life and death.

And, above all, it is still a fish of mystery. We know most of it, thanks to years of scientific study, but we do not know it all. There is still a secret life of salmon.

CHAPTER TWO

A Fish of Northern Parts

Salmon in their sea-going phases are spread across thousands of miles of ocean of the northern hemisphere. And the rest of the time run through wild coastal waters and almost uncharted rivers of the northern expanses of eastern Canada, Siberia and the Nordic lands, as well as Pacific littoral countries and the still largely unspoilt expanse of Alaska and British Columbia.

Salmon journey from those rivers and countries but then swim through national borders, including marine territorial waters, at will. The Atlantic salmon fishery, for example, is a mixed community, a markedly varied body of fish from a geographical spread of rivers that meets in common areas, if you can say that about a vast tract of ocean joined by hundreds of nautical miles of sea channel and inter-continental currents and ridges. This 'mixed stock' of fish cannot be pinpointed to individual home rivers and this concept has proven key in salmon conservation arguments, as it gives a means of countering one country's practice of harvesting stocks of salmon destined for the rivers of other countries and potentially impacting the viability of the spawning stock.

Belousiha river view, Kola Peninsula. (Where Wise Men Fish)

In the sphere of Atlantic salmon, we have known this for certain since only the 1950s. Then, in 1953, Jorgen Nielsen, a Danish fishery scientist, took a closer look at salmon that were being caught as a by-catch of cod fishers off the west coast of Greenland. This had been happening for years, but equipped with scales from the fish, Nielsen was able to see that these fish had spent two or three years living in fresh water and this ruled out an origin from Greenland's only salmon river, the Kapisigdlit. The fish must have come from elsewhere.

Then came hard proof, and the opening of a secret door to salmon understanding. In 1955, an adult salmon, stripped of its eggs, was released into the River Conon tributary, the Blackwater (Loch na Croic), in Ross-shire, in the Scottish Highlands. Just under a year later it was caught near Maniitsoq, west Greenland.

A one-off? No, it was not. In 1960, a smolt tagged in the Miramichi estuary in New Brunswick, Canada, was recaptured in exactly the same Greenland location as the Conon fish. One from each side of the pond. Other hard evidence followed.

It was the start of a process that saw the west Greenland Atlantic salmon captured and analysed and, before long, 28 genetically distinct populations of fish (20 north American groupings and 8 European), were established as being the Atlantic salmon 'races' that gathered together feeding off Greenland. Most of the European captures were from British rivers. Overall, the main origins of the north American fish caught off Greenland were: Gaspé peninsula, 23 per cent; Coastal Labrador, 21 per cent; Gulf of St Lawrence 28 per cent. Now the salmon secrets were coming thick and fast.

Researchers have found significant and useful evidence of variations in abundance based on meteorological and other data over time. At the Fourth International Atlantic Salmon Symposium held in Canada in 1992, some fascinating theories were presented by Dr Dunbar of McGill University, Montreal. He talked about a phenomenon called the Great Salinity Anomaly, which was recorded for the first time in 1968. This, he argued, had had great impacts on Atlantic Salmon numbers and he even extrapolated his arguments via a historical literature study, together with another meteorologist, Dr Thomson, to show patterns of presence and absence of salmon in their main west Greenland feeding waters:

1605-25	Salmon probably abundant.
18th century	Salmon probably scarce.
1806-12	Salmon present, perhaps abundant.
1820-50	Salmon scarce.
1890-1928	Salmon scarce.
1928-58	Salmon becoming more abundant.
1958 to the present	Salmon very abundant.

Good running water on the Nith in October when salmon run in their thousands. (HG)

Ice off Greenland marks a key feeding ground for Atlantic salmon. (Adobe Stock)

That last line is one of those moments (there are others) that bucks the trend of a reputation for scientists being doom and gloom laden. And, of course, it is interesting how a group of scientists at a symposium can shed light on the salmon's secret life.

So Dr Dunbar continued to apply the episodic abundance and scarcity model to the east coast of Canada (again Atlantic salmon), since 1910, showing a peak in 1930, followed by a decline to 1955, and then an increase in 1965. He showed the low points in this 'salmon wave' were 50 years apart, and the high points came at 35 year intervals.

Turning to the abundance of runs of Pacific salmon, many people will conjure up Discovery Channel type images of rivers so full of salmon you could almost walk across the backs of them. Add to this images of bears catching the main summer runs in Alaska, hunting salmon in the shallows, and of course swiping their share of constantly running fish over waterfalls and rapids on any number of Pacific coast rivers.

On a scientific angle, meteorologists have attributed changes in the migration route of the sockeye salmon to oscillations of El Niño, specifically the Southern Oscillation. Also on climate and the Pacific species, the International Year of the Salmon initiative fell in 2019 and part of its five-year initiative was to study ocean currents and climate change fallout across the Pacific basin. Two High Seas

Iconic across the salmon hemisphere, aurora borealis. An echo of the magnetic forces that pull salmon on their northern quest, then home. (Adobe Stock)

Expeditions focused on the Gulf of Alaska and a first pelagic ecosystem survey of the entire North Pacific Ocean got underway in planning stages to study the winter ecology of salmon in the North Pacific by late 2022, the purpose being to show the utility of such a survey to understand how increasingly extreme climate variability in the North Pacific Ocean influence salmon distribution and survival.

Other well-shared media images include chum salmon crossing a road by a flooded Skokomish River in Washington state. These, and stills and videos shot at other locations, have achieved that iconically elevated status of 'going viral', bringing salmon studies to the wider public. The Skokomish runs into a bay that is not so far south of the border with BC and Vancouver Island. It is abundant salmon country and as well as the chum or dog salmon, the river has abundant summer and early autumn runs of chinook, coho and pink (humpback) salmon, and winter steelhead (sea-run rainbow trout).

The video shows a road covered in some 8cm of water flowing rapidly right to left. And sometimes five or six hefty (2-4kg) salmon are seen crossing the road at once. Bizarrely, and dangerously for the salmon, the odd car is shown driving across. Of course, the Skokomish situation indicates how salmon will deploy full instinct mode

Perfect salmon country, this is the Gaula in Trondelag, Norway. (HG)

to find their way up their rivers, even in full flood. So sensing a good flow across a road they will follow that flow upstream.

Alternatively, salmon in high water can crop up in some unusual places. Fish are sometimes stranded when the water recedes in anywhere from an ox-bow lake on a stretch of pasture to any kind of riverside pool to a pub car park. In December 2020, for example, a pub car park on the banks of the River Fowey in Cornwall was the scene of a successful salmon rescue when local man Chris Bird spotted a stranded salmon still swimming freely in inches of water between the marked bays. An experienced angler, he had the technique to tail the fish (grip the fish in the 'wrist' of its tail with a thumb and forefinger grip) in a parking bay, swiftly transferring it in to the Fowey river proper, just the other side of the car park wall. His partner had the presence of mind to video the whole adventure – and whether they also had the foresight to line him up an ale or two at the bar as a reward for his eco-endeavour is not known.

These rescues are well-documented and the 2011 wildlife documentary, *Gaula: River of Silver and Gold* (Goez and Hamacher) shows the rescue of a Norwegian salmon of about 4kg, cut off completely from the main river in a side channel as August flood waters receded. Footage showed the painstaking and ultimately successful efforts to gather up the fish in a damp sheet and carry it back for safe release in the main river. It all made for a memorable scene in the movie. Any secret life of salmon has to log some of the more bizarre exploits like these.

The cultural importance of salmon to northern peoples and maritime based indigenous communities is clear. Cultures across the northern hemisphere celebrate and revere this anadromous fish today and have done for centuries. Images of salmon leaping up waterfalls – or being hunted by bears off waterfalls, or simply scooped in their abundance from the edges of Alaskan and Oregonian rivers in the Fall as the late summer days shorten, form part of the human experience in terms of interacting with the wild.

Salmon catch in western Canada. (Charles Hewitt)

CHAPTER THREE

From Home Rivers to a Secret Life at Sea

Salmon habitat and lifecycle mesh together from river to the sea and out into the open ocean. Then there is the return home and everything joins up again. Through the secret elements of its phases, or the bits less well documented by fishing and scientific interests, we can track the salmon's secret journey, although there are still no-go areas; riddles that have only recently been partially illuminated by what is now sustained international scientific effort, especially in terms of analysing tagging and tracking data.

The salmon's life story is a true circle of life, with the drama played out against a backdrop of life and survival, and a story bigger than itself (which as it happens, to give it an anthropomorphised twist, is the secret of happiness). It is all achieved against the current, against the flow, as it were – and against the odds. We admire the salmon's journeys over thousands of miles. The drama of its anadromous switches (from freshwater to salt and back again, involving nothing less than a chemical recalibration of its whole body as it runs up river to spawn).

So, the lifecycle of the salmon? The facts are, as they say, well documented, or 'no secrets here'. And that's the point; here is a straight account.

Let us start with the eggs laid in gravel in late autumn and winter. In early spring, the fertilised eggs, by now called eyed ova, hatch and alevins emerge from the river stones and pebbles when they have absorbed their egg sacs. These are now fry (then parr) and stay in freshwater for two or three (or four) years. However, in two species of Pacific salmon, the parr or smolt moves almost straight out to sea at this stage.

Atlantic salmon parr eat small insects like mayflies, stoneflies, caddisflies, blackflies, and riffle beetles. Sometimes they eat small amphibians and fish like sticklebacks. They feed in an omnivorous spree because there is a lot of growing to do. As always in the natural world, the bigger they get, and the speed at which they achieve that, the higher their survival chances.

Next, the parr silver up and become smolts in the spring of their second, third or even fourth year and drop down to the sea in April-June time. The migration process

Coastal waters offer a radically different environment once the salmon leaves its home river. (Adobe Stock)

allows for adjustment to a saltwater life and the smolts start feeding on plankton, sandeels, estuarial small flatfish and so on.

How long do the post-smolts (as they are known at this stage) stay in this brackish arena on their outward passage? We know that on their return from the open sea, feeding salmon swim up and down the coast with the tides, nosing into the river mouth. Especially in low summer rainfall conditions, this mix of grilse and multi-sea-winter fish will 'hang out' for days or weeks, waiting for the water and at times falling to those nets (the fixed-engine T-nets, J-nets, and drift-nets) still licensed to harvest them. The remaining Atlantic salmon driftnets of, say, the NE of England fishery, were approaching an end game and as this book went to press were losing any remaining right to compensation when, as seemed inevitable, they retired from fishing naturally.

But how long they stay there on the way out is definitely part of the secret life part of the salmon's existence. Fishers, especially on lower river pools of the salmon rivers, often catch really small grilse of less than 1kg, the size of a smallish sea-trout. Have these fish been out to the 'big water' or are they hanging out off their coastal rivers – in the case of rivers like the Orkla or Gaula in Norway, not going far from the Trondheim fjord, for example?

Cormorants and other fish-eating birds are a constant threat to salmon stocks. (Adobe Stock)

In a bid to answer this question, the UK-based Missing Salmon Project (now calling itself the Missing Salmon Alliance), promoted by the Atlantic Salmon Trust (AST), has been catching, acoustically tagging and tracking smolts descending rivers into the Moray Firth over the past two or three years. On launching, they put out a compelling statement: 'For every 100 salmon that leave Scotland's rivers for the sea, fewer than five return – a decline of nearly 70 per cent in 25 years. If this trend continues, wild Atlantic salmon will become an endangered species in our lifetime.' Whether or not you agree with the doom-laden weight of these figures, the Project/Alliance's response was to take proactive measures, catching smolts in traps, sited in places like the Feshie tributary of the River Spey, and monitoring their progress down the rivers to the sea in the spring and summer. Then – and this is vital – to try to track their progress out into coastal waters up to 200km out to sea. Mitigation of mortality was the aim and included 'understanding of the smolts' dispersal pattern in the coastal zone of their migration'. The technology tracked 800 smolts over seven river systems (the Oykel, Conon, Ness, Findhorn, Spey and Deveron) which were recorded on the receivers 15 million times.

Findhorn in early July. (HG)

Haaf net. (1) (Nith Catchment Fishery Trust)

Results from year one were interesting. The receivers, including boat-mounted receivers, used acoustic telemetry deployed across the inner Moray Firth across to Dornoch Firth. Also from Spey bay to Brora as well as a line of receivers 12 nautical miles direct north of Fraserburgh.

Fifteen million pings were received form the smolts. But in a blow to salmon lovers everywhere, higher than expected losses of 50 per cent of smolts before they reached the sea were recorded. This came as a shock to the data gatherers. It remains to be seen exactly why this is, but predation and other causes number among the potential culprits being investigated.

What was encouraging, and qualifies as a newly revealed salmon secret, was that subsequently only 15 per cent of the smolts died in the coastal waters and in fact once clear of the rivers 'the smolts moved rapidly and took advantage a range of routes out into the Firth[s] and out into the open ocean'.

It was also detected that one of the smolts travelled 200km in two weeks.

So, what happens next? Atlantic salmon travel out to Greenland and the Faroe Islands and other subarctic areas, travelling at something like 75 miles a week and showing a sense of purpose. But how do they choose their route without SatNav? And what other factors are impinging in on this critical time for salmon survival?

In his 2016 book, *Salmon and Science*, academic and fisheries researcher Derek Mills chronicles the full-on approach to the whole question of salmon survival at sea that gathered pace at an international level in the mid-1990s. He shows how there was a growing recognition by scientists that satellite tracking (in the days before we all had access to this technology via the gizmos in our smartphones) could be the answer to unravelling the secrets of what salmon did when they went out to sea.

So he shows how scientists and fishery managers started to consider the range of factors which impacted and affected or influenced salmon marine survival. These included climate, storms, ice, gyres, currents, salinity, surface temperature, migration, zooplankton, phytoplankton, water quality, pollution, movements, behaviour, forage fish, competition, growth, predation, exploitation and – last but not least – survival.

To give more background to the above, it was now emerging that what we now call global warming was creating a stink. The warming of the waters of the north-east Atlantic was, critically, changing the timings of the appearances of the above creatures.

So, while fish were meant to play a large part in the diet of post-smolts, a survey of fish in the Trondheim Fjord now showed insects making up 60 per cent of the fish stomach contents and crustaceans in 93 per cent of the stomachs. Fish remains appeared in only four per cent of the smolts/young salmon.

A three-year salmon research project funded by the British government and the Molson Family Foundation commenced in 1995. It tagged into its remit the ASF, AST, the European Space Agency and the comprehensively named Atlantic Centre for Remote Sensing of the Oceans in Nova Scotia. Included in the project was information from the European ERS-2 Earth observation satellite.

As this progressed, a couple of years on, some concerning new numbers came through on the decline of returning adult salmon and Mills underlines how the build-up or backdrop to a key two-day 1998 Freshwater Fisheries Laboratory, Pitlochry, workshop was coloured by a new urgency. The total nominal catch of Atlantic salmon had declined from a peak of over 12,700 tonnes in 1973 to 2,300 tonnes in 1997. Everyone was getting worried.

At the workshop it was Icelandic scientist Johannes Sturlaugsson who had this to say about salmon migrations (both feeding and spawning) in Icelandic coastal waters: 'The marine feeding migration of salmon is a survival combat in which the feeding conditions experienced are known to play a significant role; and the opportunistic feeding of salmon makes the best out of these circumstances.'

This really is the situation smolts find themselves in as they hit the coastal waters on an ebbing tide from their home rivers. It is war out there. They are fighting (as expert Johannes Sturlaugsson goes on to say) in an all-out competition to increase their body weight. Because they are under the existential threat of a number of marine predators, including cod, coalfish, pollack, striped bass, garfish, cormorants, shags, gannets and puffins and further out to sea or in the ocean, Greenland shark and ling. Well-shared images of the bottle-nosed dolphin crashing into adult salmon

Haaf net. (Nith Catchment Fishery Trust)

and literally taking them in mid-air may be remembered by readers in a series of photos of the dolphin-salmon hunt in the UK national press some ten years ago.

So, you get the picture – in the secret life of marine survival for seven or eight inch salmon, bulk equals survival. Johannes Sturlaugsson said 'the salmon that have for genetic or circumstantial reasons more successful feeding strategies and tactics have a head start in the race for survival.'

So it is thought that salmon do have a big secret in terms of their navigation ability – they really do have an in built system that uses the Earth's magnetic field to navigate. The post-smolts feeding as they swim, really.

In 1998 there were already a number of ways to track the post-smolts (in marine waters) including tagging and the use of existing tagged fish so archival or data storage tagged fish came into play as well as data from automated listening stations, acoustic tags and active tracking.

The secret life of salmon at sea was a little less secret and one result was some knowledge of their sea and ocean migration routes. Moving north: those from Ireland and the west coast of Scotland were hitting the Slope Current to the north west of Scotland, a current that runs north-east. Eventually these fish were segueing north-eastwards past the Wyville-Thomson Ridge, through the Faroes via the Shetlands and into the Norwegian Sea. Evidence suggested post-smolts from the east coast of Scotland (and the big four rivers of Spey, Dee, Tay and Tweed) tracked the Dooley Current, which runs from Aberdeenshire away to the Bergen direction of Norway. There is also the Fair Isle Channel, which it felt guided post-smolts via Orkney and Shetland.

Secrets about salmon behaviour are often appreciated by those with a sporting interest in salmon. To take two areas: depth and day and night time activity. When the science is observed out in the ocean it gives an added dimension. So, in terms of horizontal and vertical movements of salmon out at sea, Faroese scientists examined the vertical by using depth sensitive tags attached to fish captured at sea. Tagged fish initially made a deep dive after release to an average of 80 metres. Then they moved up in the water column to depths of 40m and some fish held at about 12m, making occasional dives as deep as 120m.

Thinking of salmon in rivers where we see fish lying at 3 metres as deep-lying fish, it is a bit of an eye-opener to think of fish regularly diving in Faroese coastal waters to 80 and sometimes 120 metres. In terms of the 'horizontal' data, fish were measured over five-day periods and periods of low swimming activity over evening hours were mixed with higher activity during full daylight hours. This tallies with in-river activity generally and in fishing terms salmon are known as an 'office hours' fish, active and proactive in the day, compared with sea trout, say, which are active at night.

More thoughts on the movements of fish centred on 'how salmon find their way back' from these ocean travels. There is still the feeling that their 'onboard compasses'

A male salmon with kyped jaw, from the Kola Peninsula's Three Rivers fishery. (Where Wise Men Fish)

Above: *Alevins. (Nith Catchment Fishery Trust)*

Below: *Annual sockeye salmon run at Brooks Falls in Katmai National park, Alaska. (Adobe Stock)*

Chinook salmon thrash up the water as they spawn on their home Canadian river. (Adobe Stock)

play a big role but that once nearer home, a number of factors come into play. These include tidal currents and the apparently random movements of salmon as they approach river estuaries, or even still far away, all part of the 'waiting game' as they wait for good conditions – including rainfall and subsequent floodwater – to run the rivers. The fish get jittery and being chased by packs of grey seals with more or less murderous intent doesn't help to calm their nerves.

I can here add my own experience, in 2005, of going out on a boat, the *Border Queen*, with one of the last sixteen licenced drift-netters out of Blyth in NE England, Steven Moss. At the time I was features editor of *Trout and Salmon* magazine and some may have thought it strange in times of heightened conservation awareness to go out netting salmon but we wanted our readers to know what was happening (almost literally) just outside the river mouth. This followed the recent buy-out of some 40 other netsmen by Orri Vigfusson's North Atlantic Salmon Fund (NASF) fundraising. We took eight grilse in his nets that day (including a 2kg grilse divided with a deft chop of a machete between photographer Peter Gathercole and me – it was fresh salmon I tasted that night) and Steven told me how he constantly saw the changes in salmon behaviour off the coast. Also how they became very active, not after river rainfall had arrived to raise the river and allow them to run, but before, as the atmospheric Atlantic front approached Blyth from

Brown trout (this is the Findhorn) predates salmon eggs and young. (HG)

the other side of the country. Interesting and a sign of another secret superpower of the salmon. Rod-fishers have noted the same changes in salmon behaviour on the approach of a weather front where a fish will take their fly which previously had ignored it, even though the rain had hardly started to fall and the river remained at the same level.

Also important in salmon finding their way back to their home rivers are 'the taste and smell of home' (homing, piloting) and orienteering. Both really are used together.

To take the last first, orienteering can involve navigation of intricate coastline (the Norwegian fjords, anyone, for variety?) as well as unique islands and scattered submarine features. Migrating birds are thought to use rivers, coastline and even artificial features like motorways and lines of electricity pylons to find their way.

In the seminal 1998 workshop at Freshwater Fisheries Laboratory, Pitlochry, Dr Bigg of the University of East Anglia had some key thoughts on climate change in the Atlantic – the subject that has had so much resonance in recent years. He said the north-east Atlantic is especially hit by 'the interactions between oceanic and atmospheric processes that can occur'. However, salmon are brilliant at adapting to these changes and that any conservation measures should remain fully flexible to take account of the salmon's own methods of survival.

The secret life of salmon has a survival purpose that may not be immediately obvious to us. Once again, the wisdom of taking a step back and not rushing in with conservation measures that could do more harm than good.

So the Pitlochry workshop brought, as always, a need for more research (into compartmentalising where ocean mortalities occurred and at what stage of salmon maturity). And here was the first real tolling of the climate change bell as regards salmon. Salmon concentrate in the north Atlantic in water temperatures of 4-8C. This situation is vulnerable to any change in ocean temperature that could affect survival. The scientists and wider salmon watching community were now starting to get the message loud and clear.

Another thing discussed and blamed for perhaps killing post-smolts at a later stage once at sea, was pollutants. Research in New Brunswick on the way 4-nonylphenol got into the food chain required examination, as well as the growing awareness of PCBs (Polychlorobiphenyls being the organic contaminant now widely blamed for contaminating marine food chains across the planet).

By the time of the Sixth International Atlantic Salmon Symposium held in Edinburgh in 2002, which detailed further tracking of post-smolts in the Norwegian Sea and Bay of Fundy, the spectre of salmon farming and in particular the problem of sea-lice infestation (especially evident at that time in the very developed fish farming

Findhorn flora. (Henry Fosbrooke)

in the Norwegian fjords) of wild stocks swimming in the area of the aquaculture cages, was raising its head.

Can we fill in some gaps in the secret life of the ocean salmon, specifically the timing of their return? After one winter at sea (one sea winter or 1SW) some return as grilse or as 2SW or 3SW fish. The sexually mature salmon stop feeding as they approach the river and this is nature's protection mechanism to stop the adult salmon denuding the river on their return. Fresh spring fish coming into the river are in the peak of condition and 'live on their fat' for three or four more months than the autumn running fish, which are beginning to colour up into their spawning colours even as they enter the river and make a direct route to 'their pool' (which may be 50 miles upstream or more) to spawn.

There are variations here – in Russia, for example, there is the Osenka, a fish that waits a whole year in estuarial or 'under ice' coastal sea water or river before spawning in the river. And then the circle is complete as the eggs are laid in the autumn. Or 'this is where we came in' as the salmon might symbolise.

So to rewind a bit (or if you like continue the circle) here is another, more jazzy, version of the salmon lifestyle story which tells us the alevin hides among the gravel as it absorbs its yolk sac, which takes a month. Then it emerges (sounds more like a secret life now?). The parr gets its own secret markings after a while – these are 'finger marks' along its side and they're quite similar to the ones trout parr also get.

Findhorn flora. (Henry Fosbrooke)

Red salmon coupling. (Adobe Stock)

The parr feed on a secret diet; flies and the larval versions of flies and tiny shrimp and snails and other crustaceans. They have to be very secret as they go about it because this is a tough time for the parr. They are preyed on by a whole menagerie of creatures. Cormorants, mergansers, goosanders and herons, of course, and many other birds including kingfishers. Mink, otters, pike, perch and eels and trout. Trout are the worst; they eat much of the parr year-class on some rivers. (This may explain the Missing Salmon Alliance reports of 50 per cent in-river losses of downstream migrating smolts.)

   Next comes a physiological change but not for two to four or even five years, but usually two. During May or June, the parr silvers up and gets a forked tail. It is about 10-15cm long and is now called a smolt and it starts to drop down the river, tail first, to the sea.

   Nature, as so often, brings its safety policy into play here because only some of a year group (or two year group) of parr make the change. Some parr turn off feeding in autumn, while their 'siblings' or 'cousins' continue. So one group skulks around in its secret existence in the river, on a subsistence diet, while the other group grow and grow and head off to sea the next spring/summer. And those on their diet finally start to eat again in March the next year and stay another year in the river.

River Findhorn salmon photographed in spring, at the end of its epic life cycle. (Henry Fosbrooke)

Of course, when they reach the river's tidal waters, the newly arrived parr are hammered by another army of predators (the Missing Salmon Alliance's 15 per cent coastal losses). Herring gulls and shags, pollack, codling, bass, congers – you get the picture. But something is now happening as the smolts rush out to richer coastal waters and eventually the open sea. They start to feed, a lot. At sea, their diet in the area around the Faroe Islands is shrimps and sandeels, while in the Greenland area they eat fish and squid. And they start to grow. A few months later they are up to 30 times bigger

Salmon Eggs. (Nith Catchment Fishery Trust)

and (if they chose this time to return home) are called grilse, weighing 1-3.5kg. Some stay out at sea longer, for two or more years and after four years might weigh 18kg.

Other marine food for the rapidly growing salmon are fish such as capelin, Atlantic herring, sand lance, barracudina and lanternfish. Crustaceans such as amphipods and euphausiids or krill are on the menu too. Cephalopods like squid and octopus are also part of the secret salmon diet along with Polychaete worms. Just before adults migrate to estuaries to begin the spawning migration, they stop eating altogether.

During that time their life under the waves off Greenland or the Faroe Islands, diving sometimes 120m below them but mostly staying 10m or so down, is still secret. At some point salmon have to return to their river to spawn because they can't spawn in the sea. It is not known whether salmon, back on the evolutionary path millions of years ago, originated in fresh or salt water and my own view is that salmon populations may have 'swung both ways', living exclusively in freshwater at times and heading out to sea at others. Not unlike *Salmo trutta* – sea trout and brown trout.

Atlantic salmon nearing spawning time. (Brad Burns)

The species, whether it is a 4oz upland river brownie or an 4kg River Tweed sea trout is the same, with only a few months of rich marine feeding (and a few hundred miles' travel) between them.

As the homecoming salmon approach their river they stop feeding. They start to lose the sharpness of their teeth and one way sport fishers can gauge the freshness of very clean salmon into a river is by their teeth. That and the presence of sea-lice (parasites that look a little like flattened tadpoles attached to their head and flanks which drop off – their long tails shortening first – within about 36 hours in freshwater).

Spring salmon entering the river any time from as early as December the previous year to May, are in tip top condition, their muscle and fat levels ready to sustain them for months of fasting. However, some fish that enter the river in, say, October or November are fast colouring up into their spawning colours and changing body shape even as they come in from the brackish estuary water. They will spawn after only a couple of weeks in the river.

River water temperature is a defining factor in salmon moving upriver. They will enter a river at 1.1 degrees C (34 degrees F) but will hang out in lower pools and be reluctant, or unable to ascend obstacles. At 4.4 degrees C (40 degrees F) salmon will more readily move upriver.

And in late spring and summer temperatures they can move fast. Upstream movement can be eight miles a day. Swimming speeds can get much faster with 14 miles a day recorded for short periods. And they travel at night – especially in low summer water conditions. This has been proven by electronic fish counters on the River North Esk. The salmon start moving at dusk and just carry on through the short midsummer nights. Tracking on Iceland's River Ellidaar has also shown salmon travelling in dense runs together, in darkness.

During summer water temperatures of 20 degrees C (68 degrees F) and upwards, salmon start to shut down their activity and this can bring on a real in-river 'secret life' stage when for sometimes weeks on end they will lie like submerged submarines quietly near the bottom of deeper runs and pools where there is the best flow of oxygen over their gills. The only thing that will barge them out of the way is a bigger fish coming up river to take the prime spot. As water levels fall, salmon can crowd into a smaller and smaller area of an upper river pool, and as more and more fish join them the party gets bigger – and more crowded. There is a great description of this phenomenon in Michael Wigan's book, *The Salmon*, talking about a pool on the River Helmsdale near his house. Wigan subsequently describes how the pod of 150-odd salmon finally moved on after some rain fell and lifted the river. When the water was clear to see into again there were no salmon to be seen. They had all moved upstream to the upper spawning reaches and redds.

Outside of the very short midsummer nights, salmon still travel at night, from say 9pm to 6am. Even if the river is so low that salmon can hardly get into the next river pool upstream, they will still move up into the shallows at the head of the pool and fly fishers swear by this time as a good one to get a take (a salmon taking their fly). Fishing the River Findhorn in extreme low water in July 2021, I was able to catch two grilse on separate days from the bottom two pools of our beat (Logie Estate), that had moved up from the rapids below at night. Seemingly, there was no other way to get a fish that week, until my fishing pal Roddie bettered me my catching three salmon from his 'Heron's secret pool' a mile upstream. But again, he caught his fish at dawn.

And now comes that most secret and mystical part of the unfolding salmon story. As the equinoctial weather systems and the October storms pass and the autumn leaves come sweeping down the rain-swollen river over the heads of the salmon, and through the water column of the pools, the salmon have changed. The hens are dark, some almost black, but otherwise retaining their 'hen salmon' shape and their bellies are full of eggs (about 700 per pound of her body weight). And the cock fish are all colours of red, with vivid patterning across their bodies and with a kype or hooked lower jaw which gives an extended head shape (the Canadians affectionately call them hook-bills). After exploring and locating an area of clean river bottom with small stones, the female swims on her side beating with her tail dislodging stones and creating a dip in the bottom of the river. She shows her readiness to mate by pressing down hard into the trough. A male fish joins her and releases milt on to the eggs she releases at the same time. The hen fish then goes a

This is the stamp of small, 2-4lb grilse that can run Scottish rivers during midsummer with this being a River Findhorn July 2021 fish. (HG)

Young salmon are monitored before return to a Scottish river. (AST)

little upstream and dislodges stones over the eggs while the male drifts below. This can be repeated several times.

There is often more than one male adult fish in attendance and 'fighting' precedes the above, with the dominant cock fish chasing any weaker fish away. What does happen, though, is that male parr, some only 10cm long, are 75 per cent likely to be sexually mature and in a precocious move they will deposit their milt in among the gravel that covers the freshly laid eggs. In fact, this is another way of the salmon making the best job of it because often in the swift river a lot of the adult male's milt is swept away in the current but the parr is able to fertilise eggs deeper down.

The fish after spawning are now seriously weakened and thin. It is really not a great number of fish that survive in the freezing winter pools. Of those that do, the kelts recover a little in the river and the species again has a way of staggering their return to the sea in the early and late spring, where some do survive and quickly recondition with rich sea feeding.

As the precious eggs nestle in the cool gravel through the freezing winter months of January and February, the salmon's secret life has come full circle.

CHAPTER FOUR

Controversial Secrets: Fish Farming and Hatcheries

It is no secret that an adult existence of 18 months duration, held exclusively within the pen walls of a salmon farm, is no life for a salmon. Or for the 80,000 others in the same cage together.

It is depressing also for people who engage with and appreciate wild salmon and a wild ecosystem. Or for anyone, really. So where did it all go wrong? *Why* did it all go wrong?

The truth is that *Salmo salar* is so adaptable that it actually adapts well to being farmed, or aquaculture as it is called. When a Norwegian entrepreneur first put salmon smolts, in a holistic, uncrowded way, into a sea pen in one of the fjords and noted how easily smolts grew on and bulked on weight, they knew they were on to something.

This seminal moment occurred only half a century ago. Norway, with its population of 4.5 million enriched by the discovery and extraction of oil, from the 1960s, and propelled into becoming one of the richest countries in the world had discovered another source of abundance. So money, and jobs, or rather the promise of jobs, brought salmon farming into being, with a more worthy reason being the imperative to feed the planet in a sustainable way. Perhaps some Norwegians, a little ahead of their time, were thinking that too. But was this a sustainable way to proceed?

The extraordinary thing about salmon farming is how exponential its growth has been, not only in Norway and Scotland, but in Chile and with a big industry in Canada. Norwegian-owned aquaculture operations have led the way, and have a financial finger – or whole hand – in many other countries' salmon farming exploits. In Scotland, whose salmon farming industry and successive governments up to the current one were wooed by, and submitted to, Norwegian outfits, there was a locking on to the 'jobs' promise. And there are jobs in aquaculture, yes, and in crucial low population density areas like the Highlands and Islands, but what about the hit to the rod fishing industry in Scotland, where hotels, gillies, fishing estate staff and local economies have been impacted since wild salmon catches crashed over the past 12 years?

A 7lb River Cains, New Brunswick, salmon caught in October is put in an in-river hatchery net. (HG)

Even just in Scotland the decline is blamed on climate change but there are concerns too about the potential impact of sea lice as wild salmon and sea trout swim past fish farms where 'clouds' of sea-lice are prevalent, washed by treatments off the caged fish. The total reported catch through rod fishing in Scotland was 37,196 for 2018, which was just 67 per cent of the previous five-year average total, and the alliance of salmon fishery boards said catches of the fish are at their lowest levels since 1952. It is important also to say that an estimated 93 per cent of this rod catch was returned to the water, so the rod-fishers were doing their bit.

In the 1960s and 1970s, salmon farming was seen as a good development. What was not to like in having huge numbers of salmon produced and the price of salmon fall to low levels that would take all pressure off the wild salmon resource? No more netting, no more illegal fishing, the end of poaching. It was a naive view, even then, but you could see what, without the benefit of hindsight, they were thinking.

A vast quantity of salmon has been farmed – a figure of 70,000 tons was quoted in 2013. The Atlantic salmon is the number one species for aquaculture and farming has proliferated across the beautiful but fragile Atlantic facing environments of the west coasts of Scotland and Ireland, not to mention Canada.

Globally the pressure to farm fish is intense in key countries which also include the Faroe Islands (Atlantic salmon are also farmed in small scale operations in Russia and Tasmania). Salmon aquaculture at the start of the second decade of the twenty-first century brought turnover of US$10 billion annually. Farmed salmon forms the bulk of salmon consumed in the United States and Europe. But – and there is a but – these caged, or penned, fish are fed pellets produced from catching other wild fish and sea creatures. Salmon grown on in sea cages require more fish than they generate as food. So on a dry weight basis, 3kg and over of wild-caught fish are needed to produce 1kg of salmon. This means that as salmon aquaculture expands, ever more forage fish is required from marine ecosystems. There is a knock-on effect and in this case it is the risk to wild predator fish which also need forage fish in the chain for their own survival.

More sobering, as if this isn't already starting to look a bit grim, is that the sea cage salmon is dosed with medicines and the pens allow parasites to get in and let thousands of tonnes of waste into the once pristine eco sites such as west coast of Scotland sea lochs where they are based. The waste just tumbles down through 30m of water and 80,000 salmons' faeces just pile up on the bottom of the entrance to some Scottish (or Norwegian, or Faroese or Canadian) lake. Annually, some 9.5 million salmon die in the salmon farms, about 20 per cent of the total, often in big events, and 'morts' (mortalities) constantly have to be removed from the cages. Then storms come in and make holes in the cages out of which thousands of salmon escape. You might think it good for them to get away from that wretched life but it's potential bad news for salmon overall.

In this way, the secret life of salmon has a dark secret to hide and it is this use of the ecologically finely balanced and primary species, *Salmo salar*, for aquaculture. And in particular the trapping of Atlantic salmon, counter to its instincts and species requirements in terms of migration and freshwater spawning, in an intensely crowded sea pen.

So what about that nice pink, through to deep red colour, in some Pacific salmon, of salmon flesh? Wild salmon get carotenoids such as astaxanthin, from their natural shellfish and krill-based diet. In a reckless policy, most farmed salmon receive astaxanthin and canthaxanthin in their feed to give that appetising pinky glow to farmed salmon. However farmed fish also get their pink flesh in part from feed that comes from sandeels and other precious feed fish hoovered up from the oceans. And that is why efforts to stop this eco vandalism and serious impact on marine food chains is to use other feeds, notably soy-based pellets.

The colour of the salmon can change and there is an impact on the omega-3 fatty acid content of the salmon. While a yeast-based coproduct of bioethanol production has been tried as salmon food, and has mooted growth properties (the caged salmon grow faster but perhaps not as fast as when fed on real fish), a more alternative food source being investigated is seaweed, which is stuffed with essential minerals and vitamins.

Clearly, there is a problem with salmon farming. On the one hand, fish farming generally, and salmon farming in particular, is held up as the key to feeding the world. What could be more ecologically sound, more sensible, more right in farming terms? And yet the charge is that salmon farming damages the seas, the inland coastal waters with fish waste and faeces, which deluges the sea bottom for hectares around the tethered cages, and with the impact of sea lice. These parasites are detached from farmed fish by chemical treatment, then float freely before, it is thought by many, reattaching themselves to smolts and post-smolts, grilse and young sea trout, killing them via sores, infection and tissue wastage, even as they travel north and south along coastal stretches heading for the paths to their ocean feeding grounds.

Sea trout have a secret life too – or did have before the farms came along and their numbers have suffered as much, if not more, than salmon. The west coast of Scotland sea trout fishery has been especially badly hit. It was once prolific on many rivers; now less so. This also applies to the rivers of Wales known for brown trout and Irish sea trout runs.

Supermarkets and other retail outlets, especially those reputed to have an 'honest relationship with their customers' have come under fire for selling farmed salmon. So often this is promoted as healthy food, high in Omega 3, other essential oils and protein, and it is. But a demand is on to stop labelling unsustainable, factory farmed salmon as responsibly sourced. Some honesty is asked for about the impacts that more information on these products would have to 'empower customers to make good choices'. The supermarkets present the farmed fish with wild looking wholesome imagery and yet is it sustainably sourced as the packaging claims? This does not stand up. Smaller species like sandeels, capelin and herring are swept up by trawlers, in order to make fishmeal to grow on the salmon so the scale of aquaculture's impact is enormous. So farmed salmon, once considered a luxury, now forms one million meals eaten in the UK every day. It is one of the UK's biggest food exports and said to be worth more than £1bn a year to the economy. M&S, Waitrose and Sainsbury's are upbeat about this 'delicious, healthy food'.

It is perhaps time, if not actually to stop outlets selling tens of thousands of tonnes of farmed salmon which scientists warn is causing existential damage to marine ecosystems, then to at least tell the truth about its provenance. To say this is a 'responsible' alternative to wild fish is flawed in these terms. So ecosystems as distant as South America and Africa suffer. The North Atlantic and Scotland's and Norway's world heritage western coastline is under pollution threat on several different levels, thanks to the farms.

Norwegian sea cages at Mowi Rakkenes. (Manfred Raguse)

Treatments for parasites kill and hurt fish both in the aquaculture sea cages, and the effects spread outside the cages. It's not just sea-lice, there are other parasites being transferred to those wild fish which keep close to coastal marine pathways. And the sea cages.

Another big threat to wild stocks of Atlantic and other salmon is the interbreeding of escaped farmed fish with wild stocks. This has been happening for years and continues to do so as farmed fish run the rivers. Farmed salmon (in Europe overwhelmingly Atlantic salmon although some coho salmon is farmed) and wild Atlantic salmon have been proven to breed together in rivers. Millions of salmon escape their sea pens in storms and mix with wild fish.

Farmed salmon have been reported running many Norwegian rivers, including the Alta and the Namsen. Many rod fishers have caught farmed fish (including yours truly, with a farmed fish of 7lb caught in July on the middle river Nith five years ago - see picture here). A spokesman for Scottish fishermen said, 'there is not one river from the Mull of Kintyre to Cape Wrath that has not been affected by farm escapees.'

Storms and sea accidents lead to regular escapes of tens of thousands of farmed salmon. While some run pretty much direct up local rivers, many remain at sea and 'mend' their appearance to start to resemble wild fish. Usually 'ragged', less than perfect fins are a give-away but here is the official advice, in this case from the Scottish government, which lists characteristics to look out for when identifying a farmed fish that has escaped the sea pens and been caught in the rivers: deformed or shortened fins (especially the dorsal, pectoral and tail fins); deformed or shortened gill covers (maybe only on one side); deformed or shortened snout; and heavy pigmentation (spots more numerous than are usual on wild salmon, and some below the lateral line, unlike wild fish) If a fish exhibits two or more of these characteristics it should be classed as a farmed origin fish. In addition, farmed fish may also have a higher number of dark spots, particularly below the lateral line, than would be expected on wild fish.

A Norwegian study has shown that nearly 50 per cent of the genetic material from farmed salmon was present in wild populations tested. The fear is that this genetic material of farmed fish could affect whole river tribes of salmon genetics negatively in the wild rivers and there are calls to sterilise farmed fish.

There are other means of 'farming' salmon. One of these is 'ranching', where fish are raised in hatcheries until they are old enough to become independent, then released into rivers to boost the wild salmon population. This segues into the use of hatcheries to supplement stocks on sport fishing rivers. The key point is that anyone may catch the salmon when they return to the rivers, so river authorities and non-profit groups will use such methods to boost stocks where they are in serious decline. It is not one for the multinationals as their investment will see a shared return.

In particular, hatcheries are used to augment salmon stocks where they have declined due to overfishing, the building of dams over time (especially on north American rivers), and habitat degradation. However, the impact on the gene pool of wild stocks again looms as an issue and this brings us to the crux of some of the current heated debate in the salmon world. What many who come out as anti-hatcheries fail to recognise is that hatcheries have been around since an earlier heyday in Victorian times. It was seen as good practice to stock River Thurso salmon into rivers all over Britain and indeed the world at one point. Thurso fish were famed for their shorter, deeper profile and 'shoulder' and their strains can still be recognised by river gillies and keepers on rivers where they were stocked (the Laggan on Islay was one, according to the gillie who worked at Laggan estates as he pronounced a freshly caught September salmon as 'a Thurso fish').

The Canadians in times past set about building a hatchery on every river and at least this kept gene pools intact. People can get too obsessed with the damage caused by mixing gene pools and how does anyone think salmon rivers got repopulated after the last Ice Age, save by salmon wandering to new rivers?

A modern example is the great operation run by the Miramichi Salmon Association. Unlike in aquaculture the efforts to re-stock, enhance and even introduce into virgin rivers, salmon runs go across the five species of Pacific and Atlantic salmon. It comes as no surprise to see the well-organised Miramichi river system in New Brunswick,

Canada, again pushing the proactive argument for salmon hatcheries. But it's not always an easy ride from historic, bolted-on hatcheries to the broad blue sunlit uplands of an eco future via salmon conservation. In New Brunswick, the federal government announced the closure of the Salmonid Enhancement Centre in 1997. It was only when biologist at the centre, Mark Hambrook, joined forces with the Miramichi Salmon Association to segue the hatchery into the Miramichi Salmon Conservation Centre, that this new and vital operation was conceived and actioned.

Fishery Director Jim Henderson's excellent hatchery on the River Nith in SW Scotland is another great example. For many years, Henderson has been supplementing the Nith's spawning stock output there. Jim is responsible for implementing the Board's strategy for fisheries management throughout the Nith catchment. Central to the Nith is its Fishery Management Plan and like many other rivers, the Nith has particular areas for focus: the mitigation of the effects of coal mining and other industrial activity; the prevention of poaching; the removal of barriers to migrating fish; the improvement of water quality and the riparian habitat; the maintenance of an appropriate hatchery programme; the encouragement of rod-fishers to 'catch and release'; the acquisition of licenses to permit the destruction of predators; the control and/or destruction of invasive non-native species; the support of science that studies British migratory fish; the exercise of influence on the Scottish Government with regard to fisheries and river management policy.

Of all of these, the maintenance of the hatchery programme is key. The young salmon swim in the main stems of rivers and bigger tributaries whereas sea trout can be found higher up tributaries and in small feeder burns. (Sea trout tend to collect in low nutrient catchments.)

All salmonids need spawning substrate – gravel in cool, clean, well oxygenated water; fry habitat – streamy bits with food and cover from cormorants. Parr need larger boulders, roots and vegetation for cover. Parr are very territorial so good cover means more eating and better growth; parr, smolts and adults all require pools to rest in on their migrations up and down stream.

So what emerges is that bankside vegetation is key and the planting of buffer zones of trees and shrubs has been a focus. The new cover soon provides food, shelter and shade in summer. Leaf litter provides food for fly life and reduces siltation and diffuse pollution from entering the watercourse. Also on the Nith, high water quality is a big priority for the fishery management team. Poor water impacts of industrial and agricultural pollution, acidification, agal bloom, droughts, and abstraction are constantly monitored. Over the last 20 years, over 77 kilometres of riverbank have been fenced throughout the Nith catchment. Habitat enhancement is carried out on an annual basis. In the work carried out by fishery managers, many kilometres of riverbank have been fenced as a result of the Scottish Rural Development Programme and Nith District Salmon Fishery Board.

This escaped farm salmon was caught in the River Nith in southwest Scotland in July. (HG)

It is important to show the sort of detail that goes into protecting the secret lives of the young salmon – and the adults – and the Nith is known to be an especially well run river. There are many other schemes working hard to turn things around – the Nith has just been proposed as a Category 3 river or one needing the most intensive conservation and care, as well as total catch and release of all salmon by fishers, again for 2022, but the eco-trajectory is looking so much better these days.

The Nith has good data for showing worthwhile returns of hatchery bred fish. Unfortunately, other rivers have a much lower hit rate. From 2010-15, releases of Atlantic salmon smolts raised in hatcheries to supplement natural production in the streams of the Gulf of Maine rivers in the US, resulted in adult spawning returns of only 0.08-0.71 per cent. This dishearteningly low return rate is the result of numerous factors including high mortality in the river from downstream passage barriers and low marine survival. The difficult upstream and downstream impact of any kind of barrier in the river will be looked at again in a later chapter.

But to finish with hatcheries on a high note, perhaps the greatest hatchery success of them all was that achieved by the late Peter Gray. Born in 1941, Peter grew up on the banks of the River Tyne. He was the grandson of a gillie on the Tweed, and soon got into fishing. Gray began fishing for brown trout but quickly discovered salmon and one thing led to another on a fish-poor Tyne in those days – until that is he started his stellar 27-year career managing Kielder Hatchery. There, Gray, almost single-handedly, brought the Tyne salmon stocks back to health and its current status as one of England's finest rivers. He went on to serve as a consultant and worked with salmon restoration programmes across the world. Salmon returns increased on the River Tyne from 724 to over 13,000 adults. His innovative rearing conditions in his unique streamside hatchery led to strong salmon prepared for survival during their epic oceanic journey. Gray, it is said, would work 90-hour weeks devoted to the welfare of his hatchery fish, growing on his fast-growing smolts and, craning over the tanks supported by his elbows, feeding them by hand high-protein booster food like chopped liver, among other things.

His approach was changing years of wasted effort in salmon conservation programmes across the world. He recognised that his parr and smolts had to be fit for the job in hand. To survive in the maelstrom of river and estuary they had to be bigger and more competitive. So he encouraged electric fly traps that deposited dying flies to the surface of hatchery pools, which the salmon parr would compete to take from the surface, rather than just being fed pellets. Here was an inspired innovation that happened through observation and accident, like some of the best scientific breakthroughs.

Crucially, he kept his young salmon longer in his tanks, until they were 11 months old, and introduced them in the autumn. The great benefit here was that the fry and parr's major in-river predator, brown trout, were more occupied with spawning activity than hammering the young fish. Gray also cooled down the water early in the life of the alevins, extending greatly the 'yolk sac' period.

By the time his little salmon went wild into the river, they were not so wild. They were lean, mean survival machines!

Truly the secret life of Gray's Kielder hatchery fish and broodstock and, frankly, methods, was something extraordinary. The pragmatic and charismatic Gray, who often came up against fishery establishment practices that held him back, started his programme in earnest in 1978 and his first problem was sourcing the parent fish.

His solution was radical. He struck a deal with the drift-net boats out in the coastal north-east of England waters. There was a good chance many of these were 'Tyne' fish but many were destined for the Tweed and other rivers. I have been out in one of those boats, the *Border Queen*, in 2005 and I witnessed how salmon would smash into the gill nets and be pulled up into the boats, without ceremony before they were taken by grey seals in attendance, and it is hard to see how they sourced undamaged fish, but they did. These were brought ashore in saltwater tanks and in due course ended up in the Kielder hatchery where Gray gradually introduced them to fresh water and they joined the cohorts of his broodstock.

Just to give a taste of the salmon abundance that Peter Gray was chasing, nearly 130,000 salmon were taken in the 1873 netting season, and a respectable 4,000 were taken by rod fishers in the year 1922, for example, before things went bad and the long, 18-mile Tyne estuary became polluted and virtually impassable to salmon.

So at the start Gray secured 50,000 'Tyne' salmon eggs using his drift-net broodstock, but he sourced another 350,000 from east and west Scotland. This made what was called a 'genetic omelette' and that glorious past from the Tyne was becoming a salmon renaissance.

The secret life of Gray's salmon became very public and Gray was feted around the world and rightly so because the Tyne, once nearly dead to salmon, was full of thousands of returning fish. There is no doubt he ploughed his own furrow, and there were whispers that he encountered some prejudice along the way. Yes, some questioned the unorthodox style of his methods. But not many.

In Pacific salmon countries, there has been a move against supplemental fish planting in favour of harvest controls, and habitat improvement and protection. However, this should be seen against the backdrop of hatchery activity on a vast scale on Pacific salmon rivers. The numbers go like this; three billion chum salmon, at least a billion pink salmon and 250 million sockeye and chinook are 'stocked'. It was in the 1970s (and in Alaska in a programme run by the private non-profit corporations, or PNP) that it was decided a bigger hatchery effort was required out west, following a worrying dip in salmon returns. So although hatcheries had been part of Pacific salmon management since the nineteenth century, a big new push to restock rivers started.

In 1968, 700 million young salmon were released, and by 1982 four billion were put into rivers. In the last three decades some five billion young salmon a year have been stocked. It is thought that ten young are released for every adult salmon caught.

In the case of ranching, fish are raised in hatcheries until they become independent young adult salmon, then are released into rivers to boost the wild salmon population. Another development from this is an even more 'direct' way of beefing up world salmon stocks, called ocean ranching, under development in Alaska, and across the border in Canada, in British Columbia. In a strategy that exploits the salmon's ability and wish to return to its embryonic haunts, the little salmon, at smolt stage, are released into the ocean well away from any wild salmon rivers.

As adult fish one or more years later, they return to spawn, but they return to that same release area, and here they are monitored by scientists and caught by designated fishing methods, perhaps somewhat bewildered in genetic imperative terms by the lack of 'a river' to ascend.

With the ever-intense bid to boost commercially profitable and very large scale sockeye salmon numbers, and other species, the salmon world eyes the largest sockeye salmon fishery in the world, with some 46 per cent of the average global population of wild 'reds'. In two recent decades, the annual average inshore run of sockeye salmon in Bristol Bay was 37.5 million fish.

There are lessons here that apply to Pacific and Atlantic salmon. In natural habitats, the survival rate of wild salmon eggs to become fry or parr of 4cm long is about 10 per cent. Predation, climate considerations, disease and food supply make up the rest. In a hatchery, however, over 90 per cent of eggs grow into fry. There are hundreds of hatcheries up Canadian and west coast US rivers and on those of the eastern Pacific.

Chum and pink salmon show a big difference from Atlantic salmon, to take one example, in this way. Although the dogs and the humpbacks are arguably less valuable salmon species, the hatcheries concentrate on them because they go to sea when they are just a few days or weeks old, whereas the other species need up to 18 months in fresh water. They are growing bigger but require a much bigger hatchery operation and numbers of water tanks. The result is more young salmon to ranch or introduce into Pacific waters.

How is all this known? Here is another secret. Hatchery salmon (predominantly pink and chum) are often thermally marked during their egg stage by changing the water temperature surrounding them by a few degrees. It gives them a tiny but distinct pattern on the microscopic earbone, or 'otolith' which can be examined at a later time when the returning adult fish is caught – three years later or so. In this way, the numbers of hatchery fish as part of the total salmon catch removes any guesswork. In 2018, for example, 34 per cent of the Alaskan salmon catch was not wild but grown on in hatchery tanks.

Controversy will always surround hatchery fish. Returning adults of hatchery origin will try to seek out their home river, but many end up swimming up another river to spawn. This puts them up against the wild fish but it can be said this is part of a natural process. It is notable that, back in the UK, when salmon started to run the Mersey again, as it were by natural accident, they were found to contain DNA from 37 different rivers.

Peter Gray (controversial to the few; a hero to the many), insisted his broodstock hatchery fish were at least 6kg but this is the exception. Here was another of Gray's insights because a criticism levelled at hatchery stocking of salmon is that hatchery fish are not selectively bred for size or any other trait. This is part of the genetic diversity debate in that salmon are generally adapted in 'tribal' terms for their own river. Even within a big river system there are different tribes of fish running different tributaries. As one source put it, 'The management of hatcheries has in the past been characterised as "Just tell me how many salmon you want". This attitude has been reformed over the last 20 years to limit their impact on wild populations.'

An alternative method to hatcheries is to use spawning channels which are usually sited next to existing salmon streams. These artificial water channels have concrete

Norwegian fjord sited salmon pens in Trondelag, Norway. (Manfred Raguse)

or other 'banks' and bottomed out with 'rocks' and gravel to recreate real river conditions. Importantly, water from the natural or original stream is piped into the top of the spawning channel, sometimes via a header pond, to settle out sediment. Unfortunately, it therefore has a 'water abstraction' downside to the river, but water is returned downstream to the natural flow.

Spawning success is often much better in these new age channels than in the 'mother' streams due to the control of floods, which in some years can wash out the natural redds (spawning areas where the salmon lays her eggs). Because of the lack of floods, spawning channels must sometimes be cleaned out to remove accumulated sediment – in other words they can get clogged up which is the last thing a spawning stream needs. That said, those same floods that damage the home stream can also clean the natural redds, which is a necessary and natural process.

On another positive note, stocking policies and hatcheries, a centuries-old ploy in salmon fishery management, has strongly improved in recent decades. This is largely due to a much better knowledge of one of the biggest salmon secrets; genetics. Advances in water purity and environmental awareness in general have produced clear, measurable and noticeable results. Add reinstatement of degraded rivers and a fight back against agricultural water abstraction, and the future looks brighter.

CHAPTER FIVE

Barometer of the Ecosystem

The ecological importance of salmon is key, as a measure of habitat degradation, and as a totem for green issues across the world's ecosystems. Salmon are tough, but they are vulnerable, a bit like planet Earth, or the human body. Or anything that is organic and valuable.

What has emerged from studies of salmon lifestyle in relation to what increasingly looks like a changing climate and environment is that, counter-intuitively, salmon can survive and thrive in unlikely conditions in the wild. In other words, they are super adaptable, in particular Atlantic salmon. This is a handy attribute in the proposed man-damaged epoch of the Anthropocene, the current geological age, viewed as the period during which human activity has been the dominant influence on climate and the environment. This is something exploited by the money men of aquaculture – the marine fish farmers who crowd *Salmo salar* into nets in their tens of thousands – to a density of 70,000 plus in individual giant sea pens – and make their lives a circular trap.

Is this what the secret life of salmon has become for the future? To be confined to the ubiquitous circular pen that is one of the classic global calling cards of modern man, with pens visible from space in myriad forms, whether as sea pens or irrigated enclosures through more or less arid parts of the Earth's surface?

You cannot separate the salmon from the ecosystem. You cannot separate the salmon's future from the future of our planet. One cannot exist without the other. Because of the spread of the anadromous king of fish across such a vast area of the Earth, and across the spheres of freshwater and marine, those who speak for the nine main species of salmon have a stake in the eco plan that world leaders are attempting to put into action. The health of salmon stocks can be a benchmark for all our futures, as much a marker for the United Nations Climate Change Conference, COP26, in November 2021 as other species and climate based indicators of the state of our planet.

There was a seminal day in the summer of 2021. It was news from the Intergovernmental Panel on Climate Change (IPCC) and it was one of those reports that was not an ordinary call to arms. It was a Landmark Report. And the warning code was red; code red for humanity. The panel used the word 'unequivocal' to

October salmon. (HG)

describe the way we have proactively raised global temperatures. A measure of global warming of 1.5C is the amount we are on track to register between 2030 and 2050 – that is a decade earlier than previously predicted.

The sobering word 'irreversible' was also used to describe effects on planet Earth due to greenhouse gas warming. To give a practical result of this – if we stay at 1.5 degrees of warming then sea levels will rise by about 50cm from what they were in the year 1900. And Canadian research scientist John Fyfe (one of 234 contributors to the nearly 4,000 word report) said this meant that by the middle of the twenty-first century there could be almost no ice left in the Arctic in the summer months.

To summarise, 'Climate change is widespread, rapid and intensifying - and it is down to us.' What the UK Met Office characterises as 'the large-scale, long-term shift

British Columbia Skeena Fishing. (Where Wise Men Fish)

in the planet's weather patterns and average temperatures', fuelled by the release of greenhouse gasses, and leading to warming temperatures, first noticed in the 1980s, was really happening. And the world remains on course to heat up to red-alert levels of danger to life.

That is the latest blunt assessment coming from the United Nations. The UN looked into the climate plans of 100 countries and concluded that we are still going the wrong way. To avoid the worst impacts of hotter conditions, said the scientists, global carbon emissions needed to be cut by 45 per cent by 2030. But this new analysis shows that those emissions are set to rise by 16 per cent during this period. That could eventually lead to a temperature rise of 2.7C (4.9F) above pre-industrial times – far above the limits set by the international community.

Autumn foliage on a Scottish salmon river. (HG)

Geological perfection for Atlantic salmon habitat represented by the Gaula river valley in Trondelag, Norway. (HG)

Within weeks and as the COP26 November 2021 date approached (with 196 countries taking part in this 'Conference of the Parties'), the opinion of climate activists was focused as never before. 'Blah blah blah,' was the withering verdict of Swedish climate activist Greta Thunberg as she summarised the commitments of world leaders on dealing with the emergency.

Greta, who pronounced forthcoming hosts, Scotland, as 'not a world leader on climate change' was not holding back. And yet she hoped to attend in Glasgow and gather with friends and young activists 'on the streets' if it was 'safe and democratic to do so', to make her point.

Carbon footprints? Global warming? In Scotland's case, even plans to develop the Cambo oil field, 125km northwest of Shetland?

Thunberg stated:

[The oil field] summarises the whole situation ... the fact that these kinds of countries who are actually hosting COP are planning to expand fossil fuel infrastructure to open up new oil fields and so on. But it is also a bit strange that we are talking about single individual oil fields, it is not just that we need to stop future expansions but we also need to scale down the existing ones if we are to have a chance of avoiding the worst consequences.'

BAROMETER OF THE ECOSYSTEM

Hammond River in southern New Brunswick. (Andrew Giffin)

As a parting shot at the September 2021 Youth4Climate pre-COP26 meet in Milan, Thunberg highlighted CO_2 emissions, even saying British prime minister Boris Johnson had 'used creative carbon accounting' in saying the UK has 'managed to reduce its CO_2 emissions by about 42 per cent on 1990 levels' at an April summit.

The IPCC's document directly referenced the eco-space of all species of salmon. It said, 'It is unequivocal that human influence has warmed the atmosphere, oceans and land.'

UN Secretary General António Guterres said, 'If we combine forces now, we can avert climate catastrophe. But … there is no time for delay and no room for excuses. I count on government leaders and all stakeholders to ensure COP26 is a success.'

What was the response from world leaders in this time of once-in-a-century COVID-19 pandemic and climate emergency? The United States announced at the United Nations 76th General Assembly in New York, that it was doubling its climate finance pledge by 2024. American President Joe Biden said, 'We must work together like never before.' He promised to increase 'climate finance' for the developing world to $10.3bn by 2024 – just over half of the European Union's pledge. The developed world had pledged to provide $100bn a year by 2020 to help poorer nations cope with climate change, but this has still not been achieved.

What has that got to do with the secret life of salmon? Everything. The oceans' warming is now accepted as the 'unequivocal' result of human influence on our planet. Rising sea temperatures have a profound effect on food species for salmon and their distribution. Journeys of 1,000 miles under European shelf and north American coastal waters to the sub-Arctic are fuelled by and existentially anchored in tracking

Lamprey. (HG)

Salmon are impacted by climate change throughout the northern hemisphere.

sand eels and capelin and smelts along the way. Migration patterns and routes can change and the timings of runs of fish into separate rivers. Once again, we get a window into a vision of wildness and also a demonstration of the strength of the salmon's ability to fight back when given the right habitat and conditions.

There is no doubt that climate change is the big factor here as it has an impact across the board and with COP26, there is yet another chance for humanity to take stock. Greta Thunberg and others argue for a systemic change that recognises a full-on emergency. Ms Thunberg still believes the conference will not lead to anything 'if we don't treat this crisis like a crisis'.

October Nith Grilse.

The change to marine ecosystems driven by human-induced conditions are physical, social, biological and environmental. The use of aquatic resources sheds light on the changes and might help to heal systemic failure.

With sea levels, for example, set to rise by 50cm from historic data by the decade 2040-50, fuelled by a global warming factor of 1.5 degrees Celsius since 1900, the evidence is clear that drivers of change are taking place faster all the time. Invasive species, dam building, regularisation of waterways, the use of pesticides and intensive agriculture, genetic modification potential impacts, mining activity and urban development all affect salmon directly or indirectly and degrade habitat and water quality.

What are the alternatives? Science can help here. How can the sometimes hidden and 'secret' lives of salmon and other fish species be fully appreciated and factored in to the solution? Because the truth is salmon can act as direct indicators in all this. Salmon can be part of the solution.

First, a caveat. Derek Mills, in his book *Salmon and Science*, talks about episodic ups and downs of salmon abundance that span decades going back hundreds of years. This gives an overview of what can sometimes be narrowed down if we are not careful to 'the last 40 years of global warming'. Fluctuating salinity in the Atlantic Ocean, El Niño cycles and developments in the Pacific (with knock on effects to the Atlantic) are all relevant here.

With greater understanding comes our best chance of helping the very survival of Pacific and Atlantic salmon, key anchors in their respective freshwater and marine

Stickleback. (HG)

Sticklebacks are prey for salmon parr. (HG)

Beloushia, Kola Peninsula. (Where Wise Men Fish, Justin Maxwell Stuart)

ecosystems. And through knowledge gained from a fellow highly evolved organism of our world, saving the planet.

A focus on salmon and the 'secret' lifestyle of salmon and the lessons to be learnt brings clarity. The wild experience of our hunting ancestors can have resonance for young people raised in towns and cities, via the direct pull of a fish on a rod (that is really what they are going for, not the fish itself) or by handling a smolt from a smolt trap on salmon conservation days. The Nith Young Anglers Club run by the Nith Catchment Fishery Trust is a good example. Worthwhile projects that have run for years in Britain get local school children out onto their local river to help with planting out salmon fry and rewilding the banks of spawning streams. These kids are thus stakeholders for their own futures as well as that of the salmon.

A history of conservation can be tracked by looking at the secret life of salmon, or the more hidden aspects of salmon populations on different rivers. So declining salmon health in Norwegian rivers, like the Tovdal, from the 1900s onwards is now known to be as a result of acidification - and the release of sulphur dioxide into the atmosphere. It also became a problem in South West Scotland and Wales and parts of East Canada.

Strangely a secret from salmon that really helped tackle another onslaught onto salmon stocks – the harnessing of rivers for hydro-electric power – was increasingly helping the development of better stocking programmes. So stocking of salmon from hatcheries which had originally been done from Victorian times with less organised knowledge of salmon genetics and holistic care for the water purity of river systems was now transformed. The results were good.

Cains River. Indicating a famous salmon spawning river in New Brunswick, Canada. (HG)

Findhorn Flora. (Henry Fosbrooke)

New Brunswick view of the South West Miramichi. (HG)

Stock enhancement together with the cleaning up of pollution in rivers was a big success from the late 1970s onwards. In the US, rivers like the Connecticut and Merrimack were cleaned up in New England. In Maine, good work was done restoring rivers like the Dennys, Aroostook and Sheepscot. In Newfoundland, the Upper Terra Nova and Great Rattling Brook were improved for salmon as was the Point Wolfe river. In England, it was the Tyne (which was to turn into a huge success story with the incredible hatchery work carried out over time by Peter Gray), and later the Torridge.

Increasingly, there was a movement to rejuvenate rivers across the rest of the Atlantic salmon world. In France, the Dordogne was tackled, after an initially excellent result in reintroducing salmon to the Bresle; in Norway, the Drammenselv and in Canada the Morell and Exploits rivers.

The 1970s had been a really low point for salmon. Acid rain, acidification, cutting down tree cover and no respect given for riparian bushes on upland spawning streams, sheep dip pollution, everything was sacrificed to what was still the post-war drive to feed a nation. In Britain, after January 1973, there was a need to access funds from the European Economic Community (EEC).

Things changed somewhat on the political front too. Strangely, it was the same EEC money-go-round which segued into the European Union (EU) in 1993 that brought some respite. Whatever your political view of the EU years, most people are agreed that the secret life of salmon got a bit better and became a bit less secretive as people were educated on the needs of salmon and for the most part their home bases of upland river systems. The fact is that the EU was good for salmon. It was good for forcing member countries to clean up their polluted waterways and beaches. The EU also played a key role in persuading member state the Republic of Ireland to submit to worldwide demands to wind up its drift net operation which was decimating the mixed Atlantic salmon stock of so many countries beyond its own shores.

A different world view started to come through in the 1990s and 2000s and with famous examples. Salmon returned to the Thames in admittedly tiny numbers but large enough to provide hope for sustainability of the runs. The same applied to the Ribble; the salmon started to lead the way as a conservational icon.

The situation today is very much based on parr and smolt survival and monitoring of salmon numbers. The study of smolts dropping down the river to the sea has become integral to the current salmon projects that closely monitor smolt survival into the marine theatre and link it to numbers of returning salmon. The results have already been surprising.

Fjord landscape. (Adobe Stock)

Above: *Fjord landscape.* (Adobe Stock)

Below: *Norway.* (HG)

Above: *Salmon country in Moray, Scotland.* (HG)

Below: *The Gaula Valley in Norway.* (HG)

Above: *The River Nith in Southwest Scotland is a classic uplands salmon river. (HG)*

Below: *Typical Kola River. Beloushia, Russia. (Where Wise Men Fish)*

Wild country around the Cains River, New Brunswick. (HG)

CHAPTER SIX

Making a Difference

Obstacles and predation block the upstream and downstream passage of salmon, and kill them, in that order. Obstacles are a big issue in North America and especially the states of Maine on the east coast (Atlantic salmon), and Oregon and Washington out west (Pacific salmon). Canada's Maritimes and British Columbia have their issues, too. Acidification is a problem in conifer forestry across parts of Britain and elsewhere; it can wipe out salmon life in upland (spawning) streams that run through those forests. Rivers in Dumfries and Galloway have suffered in some cases the loss of their spring run of salmon thanks to acidification of watercourses caused by 'tax break fuelled' conifer planting in the 1970s and 1980s. Predation includes human overfishing and poaching. It is not all 'sustainable harvest' on the wild salmon rivers.

Adult salmonids need free passage to spawning areas. That much is clear in both the Atlantic and Pacific sphere. Dams and other barriers can block their access to the headwaters' middle or upper reaches where, with some exceptions, they need to get in order to spawn and maintain the survival of their species. Where the watercourse is blocked either completely or partially, they can waste hours or days trying to get past. And this saps energy and many in the end have to spawn further downstream in less than ideal conditions.

About the only good thing to come from dams and hydroelectric schemes (which come with dams to drive the generator turbines to the detriment of salmon) is that they have promoted significant developments in salmon being artificially reared in hatcheries as a compensation measure. Artificial obstacles include dams of all varieties, but also weirs and blockages to rivers, the result of industrial needs and the building of urban areas. Forestry management, logging and deforestation with the loading of sediment into rivers is also damaging.

Fish ladders are, and always have been, a solution – and can help the adult salmon migrate up a river. They work very well. Each step of the ladder is a gradual increase in height and there are places for fish to rest. Often a ladder is put in a place that is difficult or impossible for the fish to pass any other way. Fish can struggle to find the ladder, especially in a big river – but the problem with any compensatory measure is it is unlikely it will ever give 100 per cent free access back to the salmon.

The effect on small fish is not good. Dams are more of a problem for smolts passing downstream because they can end up hitting all sorts of dangers that would not occur in a free-flowing river. In hydro schemes, small salmon and smolts can end up being sucked through the turbines if they can't find the gaps placed in the dam to aid fish migration. Roads near streams can be an additional challenge if dirt from the road messes up the spawning gravel.

The secret life of the migrating salmon (up or downstream) is dangerous, especially to the diminutive silver smolts dropping, mostly tail first, down rivers and trying to negotiate obstacles that funnel them into narrow channels, bowls of a fish ladder or anywhere where they can be picked off by fish-eating birds and animals. It applies to adults running up river of course; think of those Alaskan bears lurking at waterfalls and rapids.

There is a famous caul (weir) on the River Nith at Whitesands in the middle of Dumfries, Scotland. The ancient-looking fish pass or ladder is home to herons and gulls which pick off the smolts coming downstream, and small grilse and finnock or herling (the 'grilse' equivalent of sea-trout, i.e. smaller mature returners) running upstream at high tide, in an efficient way. In certain times of lower water, the toll of sea-going salmonids is high. It is a popular stop for Dumfries townspeople and summer visitors, across the water from the Robert Burns museum, to look down at this conduit for the entire run of the Nith's salmon population, especially from two hours before high tide, when the action can intensify. Sometimes a small silver head pops out of the water, and sometimes something bigger. Either way, you look them in the eye and urge them to get ahead of the swooping gulls and swim on upstream. It is a classic interface between man and the usually secretive salmon.

Heading north into the Highlands of Scotland, some of the first box traps put in rivers to monitor parr and smolt numbers were in the River Conon in 1959 into the mid-1960s. It was found that predation from pike, trout, goosanders and black-headed and common gulls was very high. A solution that came up on the Conon system was to take the trapped smolts and release them into the Conon below the lower dam at Tor Achilty. This produced much better numbers of surviving smolts and has been practiced for fishery purposes ever since.

Finally, on salmon ladders and jumps, and to some extent for natural waterfall jumps too, the mechanics need to be designed to include a standing wave or hydraulic jump that the salmon can use to propel themselves into their leap. So in shallow waterfall pools or below many weirs, the standing wave is too far downstream of the fall for salmon to optimise the physics of their athletic bid to strike the crest of the fall on the upward part of their kinetic trajectory.

Currently in Britain there are moves to continue the reintroduction of beavers to upland rivers. Rivers and their tributaries might benefit in some ways, but any blocking of the upstream route for fish brings its own issues for salmon.

Fish showing colour in Norway. (HG)

Dams and hydro schemes have had a massive impact globally on the salmon rivers, nowhere more so than on the NE seaboard of the United States. Michael Wigan in *The Salmon* has a devastating verdict for what happened: 'America mutilated its population of Atlantic salmon more comprehensively than any other country. Furthermore, it conducted this pogrom over a short time.' He goes on to describe how the Connecticut river, 'threading from the Canadian border through the gentle, forested, undulating scenery of New England, with 300 miles unobstructed and available to salmon, was America's richest salmon river. Even the Norse 1,000 years ago commented on the number of salmon in New England and their unusual size.'

During the 1780s, the Connecticut was still salmon-super-abundant but in 1795 the first dam was built at Hollyoke, Massachusetts, and another followed a couple of years later at Montague, which completely blocked access for salmon. Another river to go was the Merrimack which ran to the sea in Massachusetts. A big dam put in by 1822 blocked 80 per cent of the salmon access.

Head north into Maine and again there are two formerly great salmon rivers that were destroyed by dam building in the golden age of industrialisation. The Penobscot and the Kennebec rivers, however, have something a bit different; hope. As the two biggest watersheds in the state, they have more 'climate resilient habitat' than neighbouring rivers. They run far inland and have upper reaches with cooler water more suitable for Atlantic salmon to spawn in. Connecticut and Maine are getting pretty far south for salmon – almost level with the Dordogne in France.

So the drive is on to get fish, threatened now by climate change, into the upper reaches or tributaries of rivers like the Penobscot, which has to date received $60m of funding to recover salmon stocks as well as other sea-run fish. A 16 year long project has also focused on the removal of two dams such as the Veazie, breached in 2013 as part of the Penobscot restoration project and the insertion of a streamy run/channel alongside a third dam and a new fish lift at the fourth. This has opened up 2,000 miles of river habitat to returning salmon.

As for the Kennebec, the river has if anything better and more extensive salmon habitat than the Penobscot, especially in its tributary, the Sandy River. The Sandy has seen hatchery eggs planted out to boost other efforts to bring back the salmon. Although it does not sound many, the 50 or 60 returning adults in the upper parts of the Kennebec are still a big improvement on no fish at all in past years.

In places like Canada, deforestation has had a significant effect on whole ecosystems and on salmon. Forestry practices themselves can add silt to the river and degrade the river as a spawning habitat and this is especially well documented on the Pacific coast of north America.

What is also happening is that rivers are being pushed from the other end too; rising sea levels are bringing flooding and coastal erosion, both of these exacerbated by the removal or insensitive management of forest and woodland. Global warming and trees are suffering together under a deadly double tap. And the forest, which could have eased the destruction, is being razed to the ground.

Clearly, cutting down trees, and burning them, releases a lot of carbon into the atmosphere. Forests are defined by their biodiversity, a base for animals and plant life, the benefits of which have yet to be discovered. They mitigate floods and disperse coastal storms. By adding an upper layer of soil, the trees are integral to the healthy functioning of the freshwater cycle. Their most vital role however is in sequestering carbon and providing oxygen. It looks very likely that the destruction of forest contributes to ecological instability on Earth. Wildfires are also part of the problem. The environmental effects of deforestation are clear, and the disruption of the carbon cycle continues daily to devastating effect. Some people still argue that global warming is part of the cycle of life, but the most dangerous threat of deforestation is that it is connected to global warming.

As for the salmon rivers, the logging itself is for wood and paper products and, aside from mining and felling trees for cattle rearing, growing soybeans, palm oil production in Indonesia, and drilling and oil production, this accounts for another 4 billion trees cut down each year. Another direct consequence of deforestation on the salmon rivers, in North America and elsewhere, is flooding. Forests are a great absorption system for heavy rains, preventing the overflow of lakes, rivers and streams. Before they are cut down, the maritime forest canopies collect and redistribute water inland but that is lost when the trees are cut down and continental areas become more desert-like. And there is an issue also with freshwater. Where the forest floor once absorbed water and released it into rivers, via streams, gradually, the tracts of land where trees are removed shunt water straight into the rivers and nothing can sink into the water table so there is an impact with less fresh water for drinking and the rivers.

According to the World Wide Fund for Nature, deforestation accounts for 15 per cent of all greenhouse gas emissions. Clear-cutting forests – i.e. the forestry/logging practice whereby most or all trees in an area are uniformly cut down – degrades the soil which gets washed away into the streams, tributaries and rivers and silt is another problem which affects salmon spawning and fertility. So erosion on the land which shifts into the rivers causes trouble for salmon stocks. Meanwhile, eco groups have said that up to 28,000 species of wildlife could be lost in the next 25 years, because 70 per cent of the Earth's plant and animal species live in forests.

Wildfires, as happened through California and western America in 2020 and 2021, can then compound the deforestation and ecological damage. So, what can be done about industrial forestry operations to ease their effect on salmon rivers?

'Plant more trees' is not necessarily the answer and the latest thinking centres on 'rewilding'. In his 2021 book, *A Trillion Trees*, Fred Pearce argues that it is better to 'let Nature grow back trees' rather than man planting more. 'Nature works much better

Grimsa River, Iceland. Salmon Jump. (Where Wise Men Fish)

that way,' he says. 'If you stand back and don't start planting, trees will regrow. They'll reseed from a forest nearby – we actively need to give Nature room to rewild.'

The Davos Agenda at the World Economic Forum in January 2021 underlined this argument. Fred Pearce also argues that human indigenous forest communities should be in charge of the rewilding process:

> Trees keep our planet cool and breathable. They make the rain and sustain biodiversity. They are essential for nature and for us. And yet, we are cutting and burning them at such a rate that many forests are fast approaching tipping points beyond which they will simply shrivel and die. But there is still time, and there is still hope.

How does run-off from logging roads and forestry practice in general affect rivers in a Pacific salmon heartland like British Columbia? After setting out that 'over half of the world's boreal region is in Canada, which boasts a staggering 270 million hectares of boreal forest', the online eco site, The Narwhal, said that a new report from the Forest

October salmon. (HG)

Practices Board in British Columbia found that sediment running off logging roads is ending up in salmon rivers.

To summarise, the report from the forestry watchdog stated that government legislation is vague and this hole in the area's effective sediment management was risking streams like the Memekay, the Owen, the Pennask and the Woodjam:

> It's a good example of where the salmon requires legal back-up to stop a downward trend in runs of fish. Requirements of operators use only phrases like them ensuring 'forest activity does not have material adverse effect on fish passage in a fish stream'. Such requirements under BC's Forest and Range Practices Act mean it's all hard to prove and even harder to enforce. The view among activists and salmon groups is that sediment has a negative impact on fish.

Warming water temperature as the stream or river loses depth is a problem, and this is compounded by roads and culverts and other infrastructure which can cause sediment accumulation and impact movement of fish upstream and downstream. Another concern is that sediment can smother salmon eggs and reduce the abundance of plant life. It can cause 'irregular' swimming patterns by fish (observed in juvenile coho and chinook salmon) which make them vulnerable to fish-eating birds.

In the Pacific Northwest, salmon populations and the impact on them of logging have become a high-profile eco cause. Green activists assert that the Pacific salmon species are doing very poorly, even if the facts do not necessarily back this up. However, the facts do speak for themselves where numbers are available. The Tranquil River in the Clayoquot Sound area in British Columbia is now suffering steep declines in its salmon species – coho, chum, chinook, sockeye – years after being hit by significant resource extraction, mining and logging.

What can be said is that logging an area around a stream reduces the shade and nutrients available to the stream and increases the amount of silt or dirt in the water which can smother salmon eggs – or prevent them being laid in the first place. Dams can cause salmon mortality from the passage through the turbines and from birds, sited downstream, which eat the disoriented fish as they emerge below the barrier.

As for poaching on salmon rivers, the issue here is that, while poaching can be seen as a 'heritage' aspect of life, and part of the ancient rights of local people, the problem comes in its indiscriminate effect on rivers and salmon stocks. A haul of 40-odd salmon, 15 grilse and 12 sea trout was taken from one noted pool on a category 3 status river I fish, taken by a net, and the reason we know these numbers is because the poachers were caught on this occasion. The point about this is that to look at the pool you would never imagine it held so many fish. And now it does not, or did not after that particular poaching event took place 12 years ago.

Upland salmon country in Scotland's Southern Uplands. (HG)

This 2021 report of poaching comes from the website of author Brad Burns:

Unfortunately, as it always does, the hot weather and concentration of salmon in the cold water pools has brought out the poachers. People have been observed trying to snag salmon at locations where brooks enter rivers, and to run nets through pools. On August 20th guides from Country Haven were checking on the Brophy Pool when they discovered a gill net set in the broad, slower portion of the pool. The upper portion nearer the brook is pretty well protected with several anti-netting devices designed to snag and entangle any nets that contact them. The net contained two salmon and a grilse that appeared to have been dead for a couple of days. It was unclear why the net was abandoned. The word is that other lower Cains pools have been netted too. The Cains has had a good early run this year, and there are several cold water buffered pools in the lower section that are fairly remote. These make attractive targets for the poachers. There is discussion about forming a volunteer patrol to keep an eye on these pools. Let us hope that happens.

Poaching has always been there. Predation happens across the salmon world but climate change, as hinted at by Paul Rouse in his foreword, can distort predator numbers away from the natural balance of numbers and cause further problems.

So Pacifics and Atlantics, which has it easy? Pacifics have whales and bears to deal with, but Atlantics have to dodge Orcas, and they are attacked massively by grey seals, with species like mink, whether deliberately introduced or not according to geographical location, causing problems. It is worth noting that Pacifics do have a longer in-river journey to their spawning grounds, with the Salmon river in Idaho having spawning pools 1,000 miles from the sea.

An Oregon State University study, which was flagged on the Science Daily website in 2019, revealed a different picture of how and when brown bears in southwestern Alaska eat salmon. Most of these bears are hunting in smaller streams as part of their predatory rotation to catch their salmon. Small upland streams have colder water, which leads to populations of salmon that spawn much sooner in the season 'when no other populations are available to predators such as bears'. This study, and associated results, has potential consequences for how environmental impact assessments are conducted and evaluated for large projects such as the mercifully suspended mega-project of Pebble Mine in the hinterland in Alaska's Bristol Bay.

What eats Atlantic salmon? In fresh water, juveniles are eaten by a variety of fish (smallmouth bass, striped bass, Northern pike, slimy sculpin), birds (kingfisher, double-crested cormorant, mergansers, osprey, blue heron, snowy egret), and mammals (otter, mink). In the ocean, Atlantic salmon are eaten by large predatory fish like Atlantic halibut, Atlantic bluefin tuna, swordfish, and striped bass. Greenland, mako, porbeagle and other sharks are also major predators, as are seabirds such as the Northern gannet. Harp, grey and harbour seals join toothed whales such as Orcas, dolphins, and porpoises in reducing numbers significantly. Atlantic salmon are also caught for consumption by humans in targeted aboriginal or traditional First Nations fisheries.

So, finally, what can we do to save salmon? Protect their habitat, help restore streams that have been damaged by forestry and deforestation and soil erosion. Reduce instances of illegal fishing and help find ways to increase salmon survival through the weirs and blockages and dams.

CHAPTER SEVEN

The World Fish Shows the Way

Salmon species are part of a northern family. But in terms of a comparison between Atlantic and Pacific runs of salmon there is a big gap. Pre-industrial numbers of Atlantic salmon can never be properly computed but an assessment of the total Atlantic salmon catch in the period 1964-1975 is often used. It was 13,000 tons, whereas a figure for catches of all five main Pacific salmon is thought, for those years, to be thirty times greater. It gives an idea of the awesome potential of the Pacific basin and especially Alaskan waters.

The spread across the world's water and land surface carries a message. It tells us that the salmon belongs to all of us as an icon of wild nature and a symbol of conservation. And in our beleaguered ecosystemic angst, could this now be the time, more than ever, to take the salmon, this wild symbol as a signal, an already late-firing starting gun for our life or death bid to save Earth?

In taking the eco status of salmon as a marker, we can start respecting the fish across its journeys. We can stop farming the Atlantic salmon in such grotesquely intensive and often irresponsible conditions. We can look to the wild fish and stewardship of issues of smolt survival and mixing of river stock where these are relevant to survival of a species, but be bold enough to move with genetic dynamics and arguments of individual river stocks where this is inevitable. We do this knowing that fish from one river or tributary have always mixed with fish in other rivers as part of a natural plan to spread the species survival load. And remembering how, from the days of Victorians like 'father of fish studies' Frank Buckland and the nineteenth century Prussian chamberlain Max von dem Borne, the understanding of early aquaculture, ranching and farming of fish has worked with these principles. Buckland we have met already, but Borne is a fascinating character. He studied mining as a young man but focused on questions of fishing and fish farming. He is considered a pioneer in the field of fish farming and was the first to breed rainbow trout and smallmouth bass in Europe. In 1890, he introduced the crayfish, assuming that it was resistant to the crayfish plague and could replace the European crayfish, which had already been decimated. His handbook for angling fishing continues to be printed and was published in 2007 in its 19th edition, quite an achievement in itself.

Meanwhile, as one species to others in a secret life battle of survival, we are linked. Salmon and people have a bond, a calling in this regard. We will

A salmon makes epic leaps over white water on its return to its home river. (Adobe Stock)

acknowledge the stewardship role of man to grasp and answer the call to be part of the salmon magic of our local river in a proactive way. We can all make a difference and connect, inspired by a heroic species that battles heavy survival odds to thrive and inspire a brighter future for our ecological existence, a way forward for our world.

That is why our connection with salmon is so (literally) vital. You may have a personal link. You may be spotting fish in one of the great rivers of Scotland, say, the Spey, Tay, Tweed or Aberdeenshire Dee, shooting clear of the water in exuberant joy as they sense incoming rain, or to shed sea-lice – those long tailed parasites that cling on to them from the ocean (long-tailed for the first few hours, then the long tails drop away leaving a shorter bit of tail behind a 'flattened tadpole' looking 'head'). And those rivers may be local to you. Alternatively, fish jump in the autumn as the continuation of underwater jousting as numbers of salmon assert their prime spawning positions to fertilise newly laid eggs from a hen fish. Or, in the run-up to this event, simply establish mating hierarchy before the female lays her eggs in shallow depressions in the gravel river bottom.

A salmon can find its way across thousands of miles of ocean back to the river of its birth – often to the very pool it spent the first months and two or three years of its life in before dropping down the river to the sea. Salmon feed under the ice, off the Faroe Islands and Greenland and around other coastlines of the arctic north, packing on weight and bulking out their pink-hued flesh. They stop feeding several days before arriving back at their birth river and then survive in the estuaries for hours, days or even weeks before the rains come and river waters give them their passage back up the river.

What is not to be inspired by? How key and vitally linked is the status of this fish to, for instance, the International Panel on Climate Change code red light shown to humanity in the summer of 2021? In literally global terms everything is shifting, and everyone is waking up, with eco leaders of opinion and media highlighters emerging to lead the way.

Canada is a hugely important country for all the main salmon species, Atlantic in the east and Pacific in the west. That is why a feature of media commentary that stood out was discussing the eco policies of rival political parties in the build-up to Canada's September 2021 snap election. A Zoom conference event hosted by *Narwhal* and the *Globe and Mail*, in Canada two weeks before that election is a good example. Here was some sense and a true global perspective. Emma Gilchrist, editor of *Narwhal*, spoke on how in just a couple of years, since the last, 2019 election, Canadian federal political parties had now developed climate plans, including the Conservatives, for their next election push, and not before time. She said how in Canada the temperature record has been broken by 9 degrees, with temperatures to 40 degrees Celsius. This is a situation that just should not be happening with shellfish boiling to their deaths on the British Columbian coast.

Canadian salmon habitat river in June, when bright silver fish are running. (HG)

Forest fires and heatwave had hit the western Canadian province of British Columbia (and large parts of the Pacific west) in a way that gave real and tangible evidence of the link to climate change making a its fast-irreversible impact. This hopefully gives a backdrop to what can be called world salmon. And it is not just Canadian eco politics that matters for the fish; the reach carries on to the western Pacific and, in the east, the Russian Federation rivers of Siberia and the Kola Peninsula.

In the east, we can acknowledge again the notable adaptability of the Atlantic salmon. Once in the river, in, say, Scotland, Ireland or Norway, the spring running fish can survive through the summer months and into the autumn and winter on their fat and protein reserves, finally spawning in December or January and either dying or in a few cases surviving as kelts to go out to sea and do it all again. Some Atlantic salmon in the rivers of the Kola peninsula in Russia, known as osenkas, survive the whole winter under the ice, including the iced-over Barents and White seas and the iced rivers themselves. Notably, in the still prolific River Ponoi on Kola's eastern point, they spend another summer in the river alongside newer returning fish, before finally themselves spawning and making a bid back to the sea again.

In terms of the different species, here the secret life of Atlantic, chum, sockeye, chinook, silver and pink is brought into real relief and the detail is fascinating.

Guide Terry Mallin holds a 14lb salmon caught by the author on the Varzuga river, Kola Peninsula, in midsummer 2014. (HG)

Pink salmon. (Charles Hewitt)

Salmon is a common name, and the picture in the rivers and seas around the world is more nuanced, and can get complicated so we will keep things as simple as possible. But that variety is part of the joy of salmon and the whole family *Salmonellae*, which includes trout, char, whitefish and grayling.

The Atlantic Salmon is the second biggest after the chinook or king salmon. It is the longest lived, partly because a small number of Atlantic salmon survive spawning to go out to sea again and multiple spawners therefore can live longer – up to a record 13 years (which includes two or three years in the river as a juvenile). Most Atlantics live from five to eight years, one to seven years in fresh water and one to six years out to sea. The biggest recorded salmon caught on the Tweed was caught by the Earl of Home in 1841; it weighed 69 pounds. These days only a small number of fish over 30 pounds are caught each year.

Young Atlantic salmon migrate to sea every year in the spring. On the eastern side of the Atlantic, the smolt run in the Gulf of Maine begins in the middle of April and is over by the beginning of June. In yet another alteration that could be due to climate change, due to regional impacts, the smolt run is starting earlier than in the past. In truth, the timing of the smolt runs vary for all sorts of reasons. It is not just a question

The head of a King salmon. Also called Chinook, it is the biggest salmon. (Adobe Stock)

of starting later at northern latitudes – it is also a question of the natural world order dictating different timings to guard against eco disaster events that could otherwise wipe out a whole migrating year class. It is for this reason that some parr do not 'smoltify' for another year and stay in their natal river. In some sub-arctic Siberian rivers, smoltification can be delayed to the fifth year and smolts grow to the size of a trout (12oz) to survive the harsh arctic marine environment that waits for them. The secret life of smolts is integral to the salmon story.

North American Atlantic salmon migrate in the spring from the rivers where they were born and raised as pin fry, alevins and parr. They head into the Labrador Sea for their first summer, autumn, and winter. The following spring, they move to the coastal waters of Labrador and the Canadian Arctic, West Greenland, and sometimes to the waters of East Greenland. After a second winter at sea, adults from many populations migrate back to freshwater areas to reproduce, although some grilse are large and mature enough to spawn after just one winter at sea.

Then we take a big jump to the north Pacific Ocean which is home to eight further salmon species. These have a separate genus – *Oncorhynchus* – and include five main species. These are the chum or dog salmon, the sockeye, the king or chinook, the silver or coho and the pink (humpback). Like salmon everywhere they can all 'enjoy' the glare of publicity, especially at times of the lifecycle when they run back up their rivers and are so visible, but like the Atlantic salmon they have their secret lives and secret ways.

One more secretive species is actually an offshoot of Atlantic salmon. Populations of the landlocked salmon, *Salmo salar* (sebago) are found through some of the lakes of North America and northern Europe and have evolved a non-migratory life-cycle (sometimes aided or restricted by the hand of man), but are not a separate species. Genetically considered a subspecies of sea-run Atlantics, they inhabit lakes, never making that marine migration. Smaller than sea-run fish, they average between 30-50cm long.

Pacific salmon have made the leap to the southern hemisphere and chinook have been established in New Zealand and Patagonia. Coho and sockeye can be found in Patagonia. Most, as in the case of sockeye, remain in freshwater aquaculture. As always, however, there are escapes into the wild and an escapee salmon out in the ocean always has the potential to find a river, ascend it and attempt to spawn. It is what anadromous fish do.

The size of this Huchen is evident from the vintage photograph.

There are several other species which are not true salmon but have common names which refer to them as being salmon. Of those listed below, the Danube salmon or huchen is a large freshwater salmonid related to the salmon above, but others are marine fishes of the unrelated *Perciformes* order. There are two main genera and within those are trout. It is strange that the lonesome Atlantic is the only salmon in its group. But when Atlantics in their tens, if not hundreds, of thousands are pouring round the Kola Peninsula under ice strewn Barents and White Seas then it can be seen that abundance is not an alien concept to Atlantic salmon, too.

The other salmon genus, *Oncorhynchus*, contains that collective body of salmon called Pacific Salmon. They have always been far more abundant than Atlantic. The Pacific wild salmon catch averages 390,000 to an estimated figure of as much as 900,000 tonnes a year; that of wild Atlantic salmon at its peak was (note the use of the past tense) only 13,000 tonnes.

The North Pacific Anadromous Fish Commission was established in 1992 to protect salmon beyond the 200 miles zones of its members; Japan, Korea, Russia, Canada and the USA. This itself was a follow-on from the Pacific Salmon Treaty of 1985 which formalised management between Canada and the USA of the fishing of salmon which crossed each country's shores on its way to the other's rivers. The North American and Asian fishing efforts each make about half of the total catch although the majority of the more valuable species is caught by the Canadian and US fleets.

So, let us take the Pacific salmon in turn, to fill out this world salmon picture:

This Pink salmon caught from a Scottish river would have orinally been an aquaculture escapee. (Fisheries Management Scotland)

SECRET LIVES

Chinook (*Oncorhynchus tshawytscha*)

The chinook, as befits a fish with a secret life, is known by other names. Most commonly it is called the king salmon, especially in the United States. In British Columbia it is known as spring salmon. So chinooks are the big boys - the biggest of all salmon, and easily top the 14kg mark. In terms of northern reach, kings have appeared around the Mackenzie River and the far north and wild sub arctic outpost of Nunavut in the central Canadian arctic, and are reported off the California coast.

Chum (*Oncorhynchus keta*)

Also known as dog salmon or keta, chums have a special secret power. They travel far, further than any other salmon, it would seem, and they have been recorded in the eastern Pacific from north of the Mackenzie River in Canada to south of California's Sacramento River and in the western Pacific from Lena River in Siberia to the island of Kyūshū in the Sea of Japan.

Coho (*Oncorhynchus kisutch*)

Coho, also called silver salmon, enjoy a geographic sweep off the coast and in-river in Alaska and British Columbia and down the Pacific coast to California. Spawned in the freshwater creeks, cohos spend 1-2 years growing and preparing for their ocean-going adult life. It is usual for them to stay about a year and a half at sea, before returning to their home river, although like Atlantic salmon they can spend years at sea. On the down to sea journey, as in all instances of smoltification in all salmon, they prepare for the saltwater while changing physiologically and morphologically. This timeless anadromous change is, as in other species, part of their lifecycle dynamic – and the strength and flexibility of the salmon family. Most male coho return to freshwater to spawn after three years in the ocean, but some males return after only two.

Sockeye (*Oncorhynchus nerka*)

Sockeyes are known in Alaska as red salmon (cock sockeyes turn a vivid red in their spawning colours - perhaps the iconic 'Pacific salmon look' familiar from Discovery television channel-style wildlife and nature programmes). They appear in the eastern Pacific from Bathurst Inlet in the Canadian Arctic to California, and in the western Pacific from Siberia to Japan. Sockeye feed overwhelmingly on plankton they filter through gill rakers. Like other Pacific species, sockeye spawn in a single spawning life history strategy that is called semelparity. Basically this means they die after spawning in a once in a lifetime reproductive event. Sockeyes are loved by the Canadians of British Columbia, who look after this species as the national treasure and supremely valuable resource that it is.

Pink (*Oncorhynchus gorbuscha*)

Known as humpies in Alaska, pinks are found in the western Pacific from Lena River in Siberia to Korea, then turn up across the northern Pacific, and in the eastern Pacific from the Mackenzie River in Canada to California, often running shorter streams. It is the smallest of the Pacifics, with an average weight of 1.5-2kg. In 2017, a number of pink salmon were captured in nets and by rods across Britain, including fifteen in the River Tweed system. Others were caught in Germany, Denmark, Iceland, Faroese streams and rivers in Norway and Finland.

Fisheries Management Scotland put out a newsflash to anglers at the time:

> These fish were originally introduced to some Russian rivers in the 1960s, have slowly spread westwards and have now colonised some northern Norwegian rivers. These fish spawn at a different time from Atlantic salmon, have a two-year life-cycle and generally run up the rivers and spawn in summer – and often in main river channels in the lower reaches of rivers, and sometimes in upstream tributaries. Fewer fish were recorded in 2019. Due to their two-year life-cycle, juvenile fish will be derived from distinct 'odd' or 'even' years, with

the Russian/Norwegian fish being odd-year stocks. The reasons behind the unusually large numbers in 2017 remain unclear.

An odd fact is that, as Fisheries Management Scotland went on to say, numbers of pinks from the Kola Peninsula tend to be stronger and more numerous in odd years rather than even years. It would appear that the 2017 salmon originated from a particularly strong year class with good marine survival and this may explain the unusually high numbers across several countries in 2017. Liaising with Marine Scotland, NatureScot and SEPA (Scottish Environmental Protection Agency), Fisheries Management Scotland quickly produced notes on what to do if you captured or observed pink salmon in Scotland, together with an app for recording captures or sightings. The app noted the shape of tail, spots on tail, dark mouth (a 'black tongue') and image of fish in dark breeding coloration.

How can you spot the escaped pinks? You look for secretive activity of course! In this case early or unusual spawning activity in the runs and shallow connecting streams and glides on Tweed pools. There was even a request for photos or video of this spawning activity to be sent to the River Tweed Commission. 'These pink salmon have not come all the way from the Pacific!' ran the blurb:

> They were introduced to some Russian rivers around the White Sea in the 1960s and have since spread westwards and have now colonised some northern Norwegian rivers. There is also an introduced population in Newfoundland from which some rivers in Nova Scotia and Quebec have been colonised, but the most likely source of the fish that turn up in the Tweed will be northern Norway.

Personally, I would really like to catch one of these feisty little pink salmon. But then perhaps I would like to catch any salmon!

Masu or cherry (*Oncorhynchus mason*)
The cherry has a western Pacific (or from a British perspective, Far Eastern) glamour. They swim in the Pacific and in the area of Japan, Korea, and Russia.

Taiwanese or Formosan (*Oncorhynchus mason formosanus*)
This is a landlocked small salmon and leads a secret but celebrated life in central Taiwan's Chi Chia Wan Stream.

Finally, and on a slightly different tack, Danube salmon, or huchen (*Hucho hucho*), are the largest permanent freshwater salmonid species. A mighty fish to look at in pictures but I have never seen them in the flesh and they are rare, their habitat reduced by pollution and many other factors. Not a 'true' salmon? A celebrated freshwater species in its own right is the least that can be said.

RECORD BREAKERS

Weights in this section are in imperial, as this was the way the records were kept in the past. A quick rule of thumb for any reader who does not think in imperial measurements is 1 kilo equals 2.2lb, more or less! The chinook (king) is the biggest salmon, so it is perhaps understandable that the biggest one caught – in a fish trap near Petersburg Alaska in 1949, weighed 126lb. The Atlantic is the next biggest salmon. And the biggest rod-caught Atlantic weighed 79lb 2oz and was caught in the Tana River in Norway in 1928. The biggest Atlantic salmon on record was a fish of 103lb; still very big indeed, and the British rod-caught record was caught in 1922 by Georgina Ballantine on the Glendelvine beat of the River Tay. It weighed 64lb, caught on 7 October and took her two hours to bring to the boat!

Other big Atlantic salmon include: a 60lb 10oz fish from the Baltic caught in 1995; a 50lb Beauly fish caught in 1909; a reported 73lb 14oz fish caught in the Alten in 1923; also in Norway, a fish of 37.5lb from the Namsen caught in 1937; a late-season 60lb plus fish returned to Canada's Restigouche in 1988; a 55lb Vosso fish; a Namsen salmon to rival Georgina Ballantine's, caught in 1889; an Annan salmon of 50.5lb caught in 1893; an Alten salmon of 55lb caught in 1985, and from the same river a 53-pounder caught in 1968. Last but not least, there is Michael Maher's 57lb River Suir Irish rod-caught record from the year, 1874. All these fish appeared in Fred Buller's *Domesday Book of Giant Salmon* (Constable, 2007).

The biggest salmon I have seen in the flesh was a 46.5lb Atlantic salmon caught on the River Gaula in Norway in June 2011 by Arve Nilsen. The 'wrist' of the fish's tail was as substantial as the girth of a 'normal' 8lb salmon. The huge fish measured 129cm from nose to fork of tail. Once seen, never forgotten.

CHAPTER EIGHT

The Wild North, Nature and Self-Knowledge

Salmon are unique creatures, fish worth getting to know – and it can take a lifetime. Part of the mystery of salmon, and integral to any attempt to unravel its Secret Life is the way salmon runs spread out throughout the year. It is almost a year round operation, and if you include the runs of fish to the Kola Peninsula rivers underneath the frozen White Sea in winter, it is a full twelve months. In Scotland, England and Wales, the salmon fishing season follows these separate runs of fish, so the season starts for early spring fish on 15 January on the River Tay, and closes on 15 December on the River Fowey in Cornwall. In both cases, fresh silver salmon are still entering those rivers at these times.

Between these extremes are the classic seasonal runs of fish. In Britain these are spring fish from January to the end of May. These fresh, high shouldered and muscular salmon are built to endure in the river, without feeding, right through to spawning time in late autumn to winter. Summer fish come in through June, July and August. They are a mix of one sea winter grilse from small fish of one to two kilos at first in June and the start of July, with bigger grilse to four kilos running the rivers by August, bigger because of the extra sea feeding they have had. Autumn-running fish turn progressively darker, taking on the colours of spawning, so that even salmon fresh from the sea are a more burnished kind of gunmetal. Some of these autumn runners run big and distinct 'tribes' of fish like the 'Greybacks' of the Nith which used to run in October and November, carry on Nature's 'scattergun' way of protecting stocks.

Across the Atlantic, to the Miramichi in New Brunswick, there is a mid-April to mid-May run of kelts (or black salmon) down to the sea. Unlike their often diminutive eastern Atlantic cousins these are quite big and very well mended and silvering up. Then the year's supply of new fish starts with a defined June run. The early June fish are bigger but less numerous. The runs of fish pick up considerably in early July, and then from September to October comes the Fall run of Atlantic salmon. High water in high summer brings in a lot of fish that come into the river and 'hang about'. They will

The Gaula in August. (HG)

shoot up to the headwaters increasingly with further rainfall and a lift in the main river level.

Part of the allure of the salmon is that, in Britain, they are perhaps uniquely associated with beautiful upland parts of Scotland, Wales or the West Country, or NE or NW England. Western-facing steeper hillsides facing the Atlantic airflow are maybe the most enigmatic. Anyone who has walked these hills or fished these salmon rivers after rain (which they need to bring in the fish from the sea) will know and love these areas. And yet Scotland's Big Four salmon rivers, the Spey, Aberdeenshire Dee, Tay and Tweed all run on the eastern side of the country (there are geological and meteorological reasons for this, giving them greater catchment and size).

And no one is saying a Scottish west coast salmon is 'better' than an east coast salmon except perhaps in terms of fishing hut banter. They may be harder to catch, in terms of timing, because you need the correct rainfall.

Many people get interested in salmon and study them and this takes them to some of the most beautiful parts of the world. The genera name for salmon – *Salmo salar* – comes from the Latin to leap. We can watch this today with a sense of awe. For many, their first sight of a salmon will be leaping a waterfall in October or November when they run the rivers to spawn. Or they will see salmon in one of the great rivers of Scotland, say, jumping clear of the water in exuberant style as they sense rain incoming on an Atlantic weather front, or to shed sea-lice (parasites when they're fresh from the ocean). Alternatively, fish jump in the autumn as the continuation of underwater jousting as a number of cock salmon assert their prime spawning position to fertilise newly laid eggs from a hen fish. Or simply establish mating hierarchy before the female lays her eggs in shallow depressions in the gravel river bottom.

It is as if the salmon belongs to all of us as an icon of wild nature and a symbol of conservation. It is time to celebrate this beautiful creature and it calls us all to be part of the salmon magic of our local river in a proactive way.

It is also about salmon adding spice to already alluring parts of the planet. Connecting wild forestry in New Brunswick, say, with a fish that has travelled in the wild oceans over the Grand Bank and to the seas west of Newfoundland and Greenland; and back. What is really so engaging for 'salmon country watchers' is the way the secret life of the salmon is celebrated in such a public way. So those New Brunswick forests – let us take the valley of the beautiful River Cains, a tributary of the Miramichi famous for its big 'Fall' salmon. This valley is still dotted with pioneer era fishing camps, created in outdoorsman style and built from wood; all of them saluting the endeavour and sporting focus given them by a fish that has the charisma to change a landscape and a way of life.

For all the faded pioneer glamour of the Cains fishing lodges (some of them are really very humble) the Cains is a very secret valley. If any stream sums up the secret life part of the salmon's existence it would be a river like the Cains, decked out with Fall colours – the amazing late September reds, yellows, mauves and browns of the turning birch and ash leaves of New Brunswick.

Perhaps in the case of Alaska or British Columbia – already wild and beautiful – the salmon from their north Pacific feeding area are, as it were, reinforcing the wilderness; a pristine fish travelling to the secluded river of its birth, far from the crashing ocean waves, and coming full circle.

We can all make a difference and connect, inspired by a heroic species that battles seemingly insurmountable odds to inspire a brighter future for our ecological existence, a way forward for our world. The scientists have told us it is a 'code red' for humanity in terms of global warming as our planet hurtles towards 1.5 degrees Celsius of excess warming but if we adapt like the salmon and adapt in a way that shrinks our collective carbon footprint, then we have a chance. And so do the salmon.

This is not just about salmon as key species. And yet as a raison d'être for this book here is one – for all readers to make a difference.

There is a movement in Scotland to rewild. And yet many of the wild beasts of Scotland disappeared a long time ago. It is why the ecosystem is unbalanced in many

South-west Miramichi in September. (HG)

Typical upriver spawning pool. This is the Cains river in New Brunswick, Canada. (HG)

ways today. We can fret about cormorants and mergansers pressurising stocks of salmon. But there are fewer eagles in this flawed animal chain. Eagles used to eat cormorants in large numbers, helping to quell their predation of salmon.

Part of the Missing Salmon Alliance's work upriver involved counting smolts on the Feshie, a tributary of the Spey. Rewilding the Feshie was already a defined movement, fuelled by the estate's Danish owner. Some might say that being a very wealthy man he was able to 'play' with the estate as his personal project but the bigger question here is the wisdom of taking the path that increasingly looks right. That is; do nothing to the wild; the salmon will find their way back. Action, or rather inaction, can have consequences and at Feshie it might be losing local jobs in estate work. Meanwhile rewilding – woodland regeneration – managing deer densities downwards in some cases can lead to grazing pressures reduced, to allow the return of juniper, rowan and birch. Restoring abundance and diversity, in terms of flora will pull in the fauna too.

Contrary to prevailing lore, Scotland is one of the most nature-depleted countries in the world – it has lost most of its large carnivores and herbivores. It has just three per cent of its native woodland cover left. But the results are coming and capercaillie numbers are up, badgers are up, birds of prey coming back. This is a 200-year vision – cathedral thinking, as it were – it is not for us or even our children. Rewilding a place like Feshie and bringing the land around closer to a natural state than it has been for hundreds of years; is this a model for biodiversity and natural carbon sinks that should be encouraged by government policy?

And we still work to bring back salmon - and the little salmon smolts in the cool waters of the Feshie living their once secret (and now well documented) lives, because it is another piece of the jigsaw.

So we edge over to working for salmon conservation because it is worth it for us. What can we learn from salmon? Why is it important to write of and document this? Because to build up a picture of the hidden parts of the salmon's world is to better understand it.

Finally, wherever you look with salmon you find new secrets. Environmental secrets, cultural secrets. Take the Varzuga river in Russia. This place is just 50 miles south of the Arctic circle. The climate is wild and special, hosting abundant arctic wildlife in the summer and brown and black bears, which startle western visitors staying in the camps, whether they are fishing or not, by just appearing on the opposite bank as a small far-away dot. 'Those banks hold stands of Scots pine, spruce, birch, poplar and aspen, interspersed with open wetland holding bilberries, cranberries, blueberries and willow herb. Also a rich ground cover of horsetails, lichens, mosses and ferns, and fungi,' said Scottish naturalist Fred Woodward, who visited this pristine river system in 1994. And visitors are surprised to find this is a world centre for pearl mussels.

So the secret of the Varzuga salmon, and the reason the river has such staggering runs of 50,000 to 70,000 salmon ascending it annually, would seem to be – at least in great part – thanks to the beds of pearl mussels, which grow up to 12.7cm. These are abundant around Pana and other Varzuga tributaries. Woodward thought the way the mussels filter up to 30 litres of water a day is incredibly beneficial for the salmon fry. So as well as one of the best salmon rivers in the world, you get the most pristine and perfect natural ecosystem – based also on salmon and mussel symbiotic interactions – and an echo of the state European and British rivers would have been before the past 200 years of being degraded and polluted by human activity.

Conclusion

'Eventually, all things merge into one, and a river runs through it.'

Apart from the famous quotation, above, which it seems to me could describe a salmon's return from the vastness of the ocean to its home river, the American author Norman Maclean wrote of 'a sense of a river's flow and ever-changing rhythm' and it again seems to me that the river period of a salmon's life is uniquely significant to its secret existence. Hanging in the flow, adjusting by tiny movements of its fins as it glides, or 'flies' over the gravelled river bottom. Resolutely not eating, a perma-fast that lasts in some cases until death. Waiting, sometimes for weeks, sometimes for months, for its time.

Most people's experience with salmon comes according to where they live in geographical terms and in my own case the Atlantic salmon was always going to be the main species. It still seems significant that the eastern seaboard of the Atlantic Ocean is personal territory.

Atlantic salmon spend the longest time in freshwater, first the two to four years as an alevin/fry/smolt/parr in the river, then the longest time of all nine or ten salmon species in their anadromous upstream return to spawn.

That the chum and pink salmon spend the shortest time in freshwater – heading to sea when they are only an inch long – makes them no less salmon. In fact the chum (its name comes from the Chinook word, '*tzum*' for striped or blotchy, referring to its spawning markings), is one of the most numerous of salmon. And it shares this with its close relative, the pink or humpback; something to do with that super-short period of in-river exposure to predation meaning millions head straight out into the Pacific.

The search for salmon secrets goes on. Collaborative initiatives like the International Year of the Salmon (2019), which launched a five-year initiative which is ongoing, majoring on data connectivity and mobilisation. The North Pacific Anadromous Fish Commission (NPAFC) and North Atlantic Salmon Conservation Organisation (NASCO) are among the big Northern Hemisphere hitters that drive such initiatives.

'The secret life of salmon has a survival purpose that may not be immediately obvious to us.' There is a wisdom in 'taking a step back and not rushing in with

Ideal cool fast flowing water salmon habitat on the River Findhorn in the Scottish Highlands (HG)

conservation measures that could do more harm than good'. So often man can move in and try to change things and that can blow up in our faces. Take time to think and take a step back, and watch. Because there is every chance the salmon (and other species) will have the answer, and in the fight for species survival, the salmon often moves fastest – and first.

This is the truth of the earth's eco future. Less of the arrogance of humanity – more listening than hasty action. Sometimes we have to stand back and do nothing. In an echo of the forests being cut down in Oregon, Washington State and British Columbia, and leaching sediment into the salmon's spawning streams, do not cut down the southern hemisphere rainforest not knowing the secret for human survival it contains in flora and fauna not yet discovered.

The secret life of salmon is wild, authentic. We can apply science and observation and experience perhaps as an eco voluntary worker, or a scientist, or a flyfisher, imparting personal observation or helping in habitat work. In the end there is the relationship between man and salmon and that can be nurtured. Peter Coates says in his book *Salmon* (Reaktion Books, 2006):

> The fish offers much more than a scientific challenge. Relations between salmon and people in the northern hemisphere are close, extensive and long-standing: in fact, we co-evolved with them... entire communities have been embedded in salmon, giving rise to the 'salmon nation' and 'salmon people'. For tribal populations sharing its freshwater range on North America's west coast, the salmon occupied the same position as the buffalo in the lives of the Indians of the Great Plain.

Or to put it another way, we are all salmon people.

In the end, salmon connect us with the wild and the secret life of salmon is an inspiration to everyone. King Charles III, himself a fly fisher with a passion for salmon and the Patron of the Flyfishers' Club for 37 years, described in an interview in the summer of 2021 how nature has been with him as long as he can remember:

> I've always loved the countryside; I've always adored being outside all the time, because I was staying in Scotland some of the time, and in Norfolk at other times, I loved going out and exploring. For me, it's an essential part of life, to have that connection with the world outside.

A recent image of the king (when Prince of Wales), fishing the Sauce pool of the Thurso river in Caithness, Scotland, clad in traditional waders and with a traditional well-worn double handed salmon fly rod, perfectly captures the special relationship between those who fish with the river. And the salmon.

Less-known information about the often well-documented salmon is instantly engaging for those in search of answers, and a way forward for conservation. The revelation where a mostly underwater fish shatters as drops of water rainbow into our world as it jumps from the surface of a pool is truly breathtaking. So often birds, and dry-land dwelling fauna with fluffy underfur and the 'cuteness' factor win through and get the bulk of public and other funding for research but there are notable exceptions.

The current Likely Suspects Framework, billed as the flagship project of the Missing Salmon Alliance is set up with a focused, hard-evidence-based, data driven bid to link our understanding of salmon survival across the Atlantic and Pacific basins. The International Council for the Exploration of the Seas, NASCO, the Fishmongers' Company and others are on board and the mission of the LSF, as it builds its partnerships around the northern hemisphere is as follows. 'By building and sharing an agreed and coordinated dataframe structure to organise and interrogate evidence, we will not only uncover new insights but support a comprehensive international programme of new research and deliver high quality guidance to managers.'

It is easy to stick heads above parapets and get them shot down by intense media glare and this can happen in the salmon world too. But turning to the hard facts of salmon secrets of survival, the LSF is at least prepared to set out, as it were from the headwaters of a salmon river down to the oceans, the main threats to salmon survival. On its defining sea and ocean journeying, the LSF presents pointers to the natural world interactions that underline survival and even a thriving state of affairs along the way: 'Food availability, composition and quality; competitor distribution, abundance and feeding behaviour and predator distribution...'

Or as the LSF puts it:

> There are a number of candidate 'suspects' that impact on salmon survival, ranging from the obvious (e.g. being eaten by something) to the less obvious (e.g. poor feeding due to water temperature changes). We all hold opinions as to which suspects are probably responsible for salmon losses, but salmon management actions need support based on more than just strength of opinion.

This is the secret to the secret life of salmon and increasingly scientific and in-the-field evidence directs conservation effort and the associated world of data sharing. Sometimes you can look at the huge number of fisheries bodies, non-profit conservation initiatives and national governmental and non-governmental bodies, down to the absolutely crucial work done on a local, river by river, basis and there is a sense there is a lot of talk but how much action is taking place?

This would be unfair. If you look at the situation in Scotland, a historic system of private ownership of fishing rights has been proven to best protect the interests of salmon. The English model has been far less effective. In Scotland, title to salmon fishing is a heritable property right. This means control over rivers in terms of the sustainable harvest element and full on conservation effort. It is then coupled with sustained corporate effort from the 41 district salmon fishery boards, dating from two Acts of the 1860s and revised in 1986. 'Coal face' work in trapping and tagging smolts, urging good practice and care of in-river salmon during heat wave conditions, and promoting good catch-and-release attitudes and practice among sport fishers are all key. The result is that Scottish salmon fishings have, as the market says, retained their value and this indicates the health of stocks.

The health of stocks? There is currently a Scottish government assessment of the conservation status of salmon in inland waters in Scotland that is carried out each year. Each river or assessment group is assigned a grading depending on the outcome of that assessment from one to three. The 2022 assessment proposal was that 35 rivers were in the 'best' category, 37 in the middle category, and 101 rivers in category 3, where the government stated: 'Exploitation is unsustainable therefore management action, including mandatory catch and release (all methods), is required to reduce exploitation.'

Icons of salmon conservation remain, even if the late and great Orri Vigfusson is hard to replace. The point about Orri was that he was a man of action. A visionary. By persuasion, cajoling and action over years, the Goldman Environmental Prizewinner was able to raise funds and buy out salmon drift-netting operations to the west of Ireland and in NE England and in other areas. He actually made a material difference in salmon numbers surviving. Another visionary was the River Tyne hatchery guru, the late Peter Gray and there are many more. Manfred Raguse had a nice line in promoting catch-and-release for years on Norway's River Gaula at a time when it was deeply unfashionable.

The will is there to uncover the secrets of salmon in a world critically threatened by global warming. The rivers are getting cleaner for the most part but we now know it is, if you like, the air above us that holds the key in many ways.

Yes, the Secret Life of Climate Change – you didn't think we could end without bringing this one up again? A January 2021 blog post by the British Meteorological Office said the following about the previous year's weather pattern:

> 2020 was a remarkable year for our climate, with the year being the UK's third warmest, sixth wettest and eighth sunniest in the UK national series, extending back to 1884, 1862 and 1919, respectively.
>
> An attribution study looking at the temperatures for 2020 has been produced by the Met Office using peer-reviewed methods. The result is available on 'Carbon Brief'. It suggests that without human-induced climate change, a year as warm as 2020 in the UK would have a likelihood of 1.1 % (uncertainty range

0.9 % – 1.3 %), or around one year in 90. For the present day climate the likelihood estimate increases by around a factor of 50, to 56% (range 53% - 58%) suggesting an expectation that we would now expect around half of years to exceed the warmth of 2020.

The impact of climate change on total annual rainfall for the UK is less clear than for temperature. This is a consequence of the high variability in rainfall over the UK, and also because climate change is most likely to result in wetter winters and drier summers overall, change which will consequently be less obvious in an annual total. However the likelihood of years as wet as 2020 is increasing, and is expected to continue to do so through the 21st Century.

So what does this mean for the Secret Life of (in this case the Atlantic) salmon? Salmon fishers, certainly, will tell you that 'extreme' weather is either flooding them off for their week's fishing, or ruining it by reducing their river to a drought impacted stream. But remember this; for salmon, and by extension all living creatures, nature has given them a scatter gun way of coping, spreading smolt runs to the sea to different years for the same year group, for example. The Darwinian intricacy of a perfectly adapted fish can even at times mask the sheer urgency of the situation.

Meanwhile, we continue to try to unlock a way forward on climate science and even as I write this conclusion, news comes through that three scientists have been awarded the 2021 Nobel Prize in Physics for their work to understand complex systems such as the Earth's climate, and, via computer models, 'reliably predicting global warming'. Syukuro Manabe, Klaus Hasselmann and Giorgio Parisi were announced as the 2021 winners in Stockholm, with the Nobel Committee saying that Manabe and Hasselmann 'laid the foundation of our knowledge of the Earth's climate and how humanity influences it'.

The message of the 26th UN Climate Change Conference of the Parties (COP26) held in Glasgow in November 2021must not be dimmed by the appalling human and environmental tragedy of Russia's war in Ukraine, planned and brutally instigated in the year that followed, just as humanity was starting to emerge from the Covid 19 pandemic.

And finally, the secret life of salmon is also a metaphor for the still small voice of salmon. A voice of reason. The still voice that tries, along with other parts of our ecosystem to bring us to our senses.

We know what we will miss out on if we carry on down the road of degradation of our blue-green planet. Earth will not be so green and it may be more blue than we had bargained for with rising temperatures and sea levels covering the globe in more water. Whatever the trajectory for our world as we try to slow down climate change, in salmon terms the years of relative abundance will still occasionally occur. The

The author talks with salmon conservationist Manfred Raguse in Norway, 2011. (HG)

miracle of the salmon species is that they adapt, but for how long? And they cannot do it in isolation.

Increasingly we can see it very well, and we have to hope it is not too late to lessen the impact of the Anthropocene. It is still only a proposed epoch - this 'human selfishness era' - and we must hope there is still time to act and learn the lessons of the secret life of salmon and rewrite history.

Given that there is still time, and it at last seems reasonable to think there might be, then we can all play our part.

APPENDIX ONE

Seasons of Change: A Year on a Salmon River

Is there a paradox in highlighting the secret life of salmon? Does it even matter? The conservational world and fishing and the wider media are full of the significance of salmon and its importance as the 'king of fish'.

Can salmon be said to have a hidden life? Surely a fish was never so well documented, with hours of streamed and terrestrial television footage and commentary, small libraries of books and academic publications, photographic banks of stills and internet-based and alternatively sourced videos and wildlife and fishing movies are testament to the box-office draw of the salmon. Its dramatic lifestyle and specialised pull on our imaginations is undeniable. Instagram does salmon; YouTube does salmon; any day now, it will be on LinkedIn!

And yet there is a gap.

This is the secret world of the title. They say that it is a good thing to keep a secret, or keep the mystery and it is this which may yet save the nine plus species of salmon, and the wider ecosystems of the planet, for our children, their children and the generations to come.

A good way of unlocking the salmon's secrets is to examine the timings of its runs into its home river through the season. Runs of salmon vary from river to river but basically have a spring, summer and autumn run. As a starting point, the eastern Canadian, New Brunswick, River Miramichi is as good a place as any.

Here there is a mid-April to mid-May run of 'black salmon'. These are what are called kelts on some other rivers, salmon that have spawned but are 'in recovery'. The interesting thing about the Miramichi is that within the whole salmon world, along with some of the Russian rivers, there is a relatively high survival rate of kelts, hence the 'black salmon' category. In fact, these fish are actually fished for in the spring by some anglers.

With June comes the salmon run proper of clean fish from the ocean, the bigger salmon coming into the river as well, although these are fewer in number. Runs pick up in early July (they did in 2012 for example) but this can be very weather dependent. High water in the summer can make a lot of fish come into the river system and then

A Scottish October-caught fish is returned to the river to spawn. (HG)

hold in the mid and lower river, although there can be some movement back to the estuary and coastal waters.

Finally, from September to October 15 is the Fall Run – although the second date is the end to the fishing season, runs of fish will enter the river long after this.

The year on most salmon rivers comes around like this and water levels, marine conditions and tides all play on the timings of salmon runs and, remember, these are always backed by an eco-insurance policy where different timings of runs, and size of fish, are mixed up in case there is a massive climatic or other event that could otherwise wipe out a river's stock, or large parts of it for a period of years afterwards.

Part of the mystery of salmon, and integral to any attempt to unravel its Secret Life is the way salmon runs spread out throughout the year. It is almost a year round operation, and if you include the runs of fish to the Kola Peninsula rivers underneath the frozen White Sea in winter months, it is. In Scotland, England and Wales, the salmon fishing season follows these separate runs of fish, so the season starts for early spring fish on January 15 on the River Tay, and closes on December 15 on the little River Fowey in Cornwall. In both cases, fresh silver salmon are still entering those rivers at these times.

In the meantime come the waves of fish, running upriver from the sea. In UK rivers these are the 'springers', which run from January and February to the end of May. These fat, powerfully muscled salmon, with high reserves of fat, are made to survive in the river for months without outside sustenance right through to their spawning the following early winter. So-called summer fish come in after this, swimming in from the estuary through the months up to August and even into September. These are made up of one sea winter grilse from small fish of two and three pounds in June up to seven or eight-pounders which enter the rivers in late summer (more time at sea equals heavier, stronger fish). Autumn-running fish turn darker in colour and some are the muscular monsters of angling myth and legend.

Across the Atlantic, to the Miramichi in New Brunswick, there is a mid-April to mid-May run of kelts down to the sea. Unlike their often-diminutive eastern Atlantic cousins these are quite big and very well mended and silvering up. Then the year's salmon run of new fish starts with a defined June run. The early June fish are bigger but less numerous. The runs of fish pick up considerably in early July, and then from September to October comes the Fall run of Atlantic salmon. High water in high summer brings in a lot of fish that come into the river and 'hang about'. They will shoot up to the headwaters increasingly, with further rainfall and a lift in the main river level.

What really constitutes the secret life of salmon, though? How really can we quantify it, approach it, expound it? This secret life of salmon from a salmon's point of view – is that even possible? From a salmon's eyes? Probably not, because as humans,

Rainwater in July brings the first silver salmon in numbers. My father caught this 13-pounder, his first-ever salmon, on the Nith in 2009. (HG)

as hunter-gatherers who have had a relationship with salmon from the beginning, over tens and hundreds of thousands of years, over millions of years if you include our own DNA stretching back into our animal ancestors that came before the ascent of man. how can we begin to communicate something that is innate, instinctive, built deep into our DNA, our umbilical connection with wild Nature of the north of the globe and the silver fish that swim in its oceans?

We have always had this relationship of hunter and prey and yet as they say, who is in control? The first fish nosing up the rivers in January or the very early spring are the pioneers. Like our native American ancestors and cousins, we catch the first fish gently. The salmon, says folk tradition, allow themselves to be caught, in our network of nets and palisades and false bottoms and channels that, in the ancient way, catch and trap the fish on the outgoing, ebb, tide (the ingoing flood tide lifts the fish over our intricate structures) and we treat them well and gently, with respect and thanks even though we kill them for our first post winter meal. Eating the salmon flesh that tastes of the sea after months of overwintering on dried salmon – a quite different flavour.

So the secret life of salmon is also our secret life. Our secret and precious relationship with these wild fish and our planetary home. Even as modern twenty-first century humans, we as fishers pour a libation into the river, a malt whisky from the quaich, to salute and pray for the salmon (and the river). We have taken to catch and release because of that ancient respect and we now know that releasing the salmon alive back into the water feels like respect, is respect and care and love. It feels like the right thing to do – and it is.

We can ponder the secrets of the salmon that are so sought-after to us. We can see their underwater, out at sea secret times and extrapolate and extrude the bits we know so much better when we have a salmon in our arms next to a river. The secret life is said – or said internally – with a whisper of respect and shared experience of life's drama.

APPENDIX TWO

Salmon Facts and Factoids: A Cheat's Guide to the Secret Life

The natural world is full of wonders and every animal has its little secrets, if we define a secret as 'something not everyone knows'. Salmon are no different and there are lots of little facts and figures about them which are nice to trot out in conversation, to amuse and amaze.

MANY FARMED SALMON ARE DEAF

As we learned earlier in the book, many farmed salmon have the otolith in their ear thermally marked by varying the water temperature in which they grow. This is intentional and can be used to identify a fish if it escapes and is subsequently caught – it is useful to be able to track what happens in these cases. However, an unintentional side effect on the otolith is that the fast growth rates in pens can affect the development of this bone and it can therefore fail to develop properly, making the salmon unable to hear underwater sounds and also have impaired balance when swimming.

WILD SALMON'S HEARING

Assuming that a salmon is not hearing impaired, they can actually do it very well. They hear using low frequency sound waves which vibrate through the water to a row of sensory pores called lateral lines on the sides of the salmon, but it is not known quite what use they put this information to, whether it is for homing purposes or simply to avoid predation where possible.

A Norwegian 7lb August salmon showing onset of spawning hues. (HG)

FINDING THEIR WAY HOME

While it is known (by using tracking devices) that salmon prefer to return to their home rivers and pools to spawn, it isn't known for certain how they do this. It seems clear that they use the earth's magnetic field to an extent, but another school of thought is that they can navigate by smell and taste of the water. It has been proved in tests that Atlantic salmon can smell one drop of scent in an area the equivalent of ten Olympic size pools, that is 3million litres! Put more scientifically, it can detect a specific scent to a degree of one part per million. This is nowhere near as good as a dog (1 part per trillion is the average for a trained canine nose) but many times better than a human. So perhaps 'smelling their way home' is just a walk in the park for a salmon.

WHAT IF THEY GET 'LOST'?

If a salmon is unable to find its home river and pool, or unable to access it (for example an obstruction or extreme low or high water), it needs to quickly find another pool or river, or it will die. This seems harsh at first sight, but in fact is a part of natural selection and the spreading of the gene pool. Without these 'lost' salmon, weaknesses inherent in a specific population can become a threat to their survival, so a little mixing now and then is a good thing. Though of course a counter argument may be that if the mix comes to include a salmon gene for not being very good at geography, this might not be altogether helpful!

LITTLE AND LARGE

The smallest salmon is the Pink or Humpback salmon, weighing from 1.4-2.3kg although some get to over 4.5kg. The biggest is the King or Chinook salmon, and it is no secret this can grow to upwards of 45kg with a weight of 57kg recorded. Sport fishers would view a fish of 22kg plus as a fish of a lifetime and two of my fishing pals have caught kings of this size. One of these, the late Roddy Dickson of Dumfries (who shot for Scotland), showed me his mount of this awesome fish. From the sizes above, it is clear that even a 'small' salmon is still pretty hefty. For scale, the average trout you might see on a fishmonger's slab weighs in at a mere 25g, less than a quarter of the size of a small Pink.

EATING UP THE MILES

Salmon can travel over 1,600km in their journey to ocean fishing grounds. Initially there is the distance travelled as smolts from their birth river to the sea, which can be 50km in a medium Scottish river, but hundreds of kilometres in a big north American river system. It is therefore hard to be precise, but even a fish who keeps pretty 'local' will rack up a journey equivalent to travelling between Barcelona and Vienna, to choose two completely random points on the globe.

CAN ATLANTIC SALMON MATE WITH PACIFIC SALMON?

Even though they are distantly related, have fairly similar lifecycles, and behavioural patterns, Atlantic and Pacific salmon cannot interbreed. They belong to different genera and have different numbers of chromosomes as well as their geographical separation. It has been suggested that farmed Atlantic salmon could crossbreed with Pacific salmon, but so far, studies have shown that this is not possible. However ...

CAN SALMON AND TROUT CROSSBREED?

Atlantic salmon have successfully mated and produced offspring with brown trout, although this only seems to happen very rarely and in controlled (in other words, forced) conditions. In the wild, this is unlikely to happen, though in stressed environments, anything is theoretically possible. None of the Pacific salmon types can successfully crossbreed with any trout species.

PREDATORS

Salmon have many predators – including man, of course, though farming rather than hunting is replacing that side of human behaviour – and some are more successful than others. Obviously a fish the size of a salmon is outside the range of many predators, although large sea birds have been known to lift a smallish salmon from the water. It is hard to quantify the amount of salmon lost to direct predation – as opposed to sea lice infestation and other disease systems – but it has been assessed that an Orca will eat on average 25 kilos of salmon a day. Taking British Columbia as a typical example, their estimated population of Orca comes in at around 300, equating to a loss to predation of some 1,000 tonnes of salmon or, to make it easier to picture, around 100 large trucks.

Aleutian Rivers, Alaska. King salmon. (Charles Hewitt)

AND … JUMP!

A salmon can easily clear an obstacle two metres in height, as most of us have seen either on TV or in real life. This is a typical jump for an athlete on top of his or her game, but this is for a person already almost that tall. So for the salmon, in relative terms, it is like a person clearing a bungalow in one bound.

SCALES

Salmon scales can be an indicator of the life of a salmon. Atlantic salmon have rings on their scales which can tell us how many times they have returned to the river to spawn and then gone back to the sea. Each event leaves the ring – similar to the growth rings on a tree – and these are therefore really helpful indicators to researchers. In Pacific salmon, an interesting fact is from the largest to the smallest, the number of scales is the same, they simply vary in size. For researchers, the scales are a rich source of DNA and of course very easy to harvest with no harm coming to the fish. With the growing bank of DNA material, it is possible to tell the exact source of the salmon, down to the section of river where they were hatched in some cases.

It is a fact that Salmo salar caught in Russia are the same species as Eastern Canadian fish. (HG)

TELLING SALMON APART

To many people, a salmon is a salmon is a salmon but hopefully, as we reach the end of this book, the differences have been made a little clearer. But even so, when presented with a salmon (or a picture of one) how easy is it to tell one from the other, size apart?

King
A king salmon is an easy one – the inside of its mouth is an overall black, the only salmon to exhibit this feature.

Pink
A pink salmon may have a black lower jaw but the upper jaw will always be white. The tail can look very like that of a King, so this is an important way to tell the difference.

Silver
Again, the mouth is the clue. The lower jaw where the teeth emerge from the gum has a white line, very clear and looking as if it has been put there with a white pen. This is unique to this fish.

Salmon fishing in Aleutian Rivers, Alaska. (Charles Hewitt)

Chum
The mouth isn't the clue this time, but the eyes; the pupils of a chum are unusually large, making the eyes look very dark. There is also a very clear dorsal line.

Red
The red's body is remarkable torpedo shaped with a snubbed nose. Its scales are also rather more neatly aligned than many of the other salmon. In a way, in its lack of distinctive features, the red could be called a 'typical' salmon, what most people would draw if asked what a salmon looks like!

NUTRITIONAL VALUE

In a book celebrating the joy of having wild salmon in our world and the important place it plays in our ecosystem and mythology, it seems almost cruel to discuss its value as a food, but essentially, that is why we are so interested in salmon and have looked after it – with varying degrees of success – for millennia. There is a slight difference between farmed and wild salmon, with wild being higher in protein and farmed higher in healthy fats and calories. Salmon is particularly high in selenium, an important nutrient that is involved in DNA synthesis, thyroid hormone metabolism, and reproductive health. It is also a great source of omega-3 fatty acids, a type of heart-healthy fat that can decrease inflammation and support brain health. Finally, salmon is rich in vitamin B12, which is necessary for producing red blood cells and regulating the health of the central nervous system. And it is, of course, delicious!

APPENDIX THREE

The Battle of the Smolt Trackers

Smolts are the final juvenile iteration of salmon, the teenagers or young adults of the salmon world. In a high altitude, upland stream like the Avon, the main tributary of the River Spey in Scotland, they originate from eggs laid at an altitude of 640 metres. In such higher, less nutritionally rich, waters they can take longer to 'grow' to smolt stage, as long as four years; indeed, some rivers even have smolts that take five years to reach that point. We are at a stage now where we know more and more about smolts and their downstream journey from the headwaters to the sea. That's because more and more salmon conservation organisations are studying this key stage of the salmon lifecycle. In fact, this separated input of effort into smolt study forms the basis for this appendix. While researching three separate smolt programmes, it became increasingly interesting to compare these separate efforts to smolt track and therefore to make comparisons and see what could be learned. It is my hope that the conclusions drawn here will help the somewhat disparate studies of in this case, the AST, ASF and NDSFB and including also the Miramichi Salmon Association (MSA) all give good pointers of good practice, while utilising somewhat differing methodology.

The ground breaking MSA operates smolt tracking programmes into early June (2021) in which they use four wheels across the Miramichi system and combine an actual count with an estimated programme. The benefits of the MSA work are clear to be seen – and apply as they do to a huge river system that holds one quarter of all North America's Atlantic salmon run. Then there is the work of the well-funded Atlantic Salmon Federation in what they call 'tracking the Atlantic salmon's murder mystery at sea'. Despite much research, there is still a real mystery about the salmon's life once it heads from river mouth either side of the Atlantic to the feeding grounds off the Faroes and Greenland, and other places.

2021 marked the ASF's nineteenth year for tracking smolts out to sea, and the fifteenth year of tracking adult kelts out to sea, those hardy survivors returning after their in-river spawning. The ASF had the vision to embark on this project and teamed up with DFO and the tracking technology of VEMCO. In that time, ASF has implanted transmitters in 4,500 smolts in various rivers in the Canadian Maritimes and Quebec, and 550 transmitters in kelts.

The project has relied on a combination of efforts and talents. On the Restigouche River in the summer of 2021, ASF, NB DNR staff and volunteers have been assisting Gespe'gewaq Mi'gmaq Resource Council (GMRC) with catching and tagging kelts. It has been a successful collaboration. It is organised as part of the Environmental Sciences Research Fund (ESRF) programme, which is an Indigenous collaboration with DFO and NGOs through Atlantic Canada. The work aims to monitor, through tag and track, Atlantic salmon kelts and smolts on a broad scale throughout Atlantic Canada and into the North Atlantic to shed light on the secret of their migrations, focusing in particular on areas where high mortality rates are hurting salmon stocks.

The kelts in 2021 were both numerous and in good condition, and hopes were high for these recuperating fish to make the round trip back to Greenland waters. The acoustic transmitters provide identification and other information as the fish swim through lines of receivers in places like the Baie des Chaleurs, at the Strait of Belle Isle, and through a line near Port Hope Simpson, Labrador.

A second type of transmitter is the satellite pop-off transmitters that are also being used. These are attached externally to the salmon, and after a set number of months, or if a salmon dies or is eaten by a predator, the buoyant transmitter rises to the ocean surface and sends data to satellites in space. Temperature, depth, and travel details can all be retrieved. If the Atlantic salmon has died, information on temperature can help determine whether it was eaten, and if so, information on which predator species is most likely.

ASF has not just worked with the satellite tags in eastern Canada. They have also worked with NOAA, DFO and KNAPK to capture adult Atlantic salmon in Greenland waters, especially near Qaqortoq in the southwest of the country. Once again, collaboration among origins is key to pushing forward this important work of understanding Atlantic salmon movements.

Recently ASF Biologist Graham Chafe gave a 13-minute Zoom presentation on ASF's kelt tagging programme. The raw data is just the first part of this important research.

Now with many years of information, inherent patterns can be explored for details, and changes can be observed. For example, on the Miramichi, ASF's tracking data began several years before the explosion of the striped bass that have become a major predatory species, especially in the Miramichi. The information derived from the tracking programme has proven invaluable in understanding the relationship between the two populations.

Another important direction for researching has been understanding the details of the smolt migration from rivers such as the Miramichi, Restigouche and Cascapedia across the Gulf of St. Lawrence, and what physical factors are impacting this movement of millions of young Atlantic salmon.

ASF Biologist Jason Daniels recently provided insight on how temperature impacts the migration and affects how the smolts move toward the Strait of Belle Isle, the narrowing of the ocean between Labrador and northernmost Newfoundland. This understanding of the nature and dates of the migration could be absolutely vital in safeguarding that movement.

Jason Daniels' 35-minute Zoom overview of this research is yet one more example of how the Atlantic salmon tracking programme is helping us better understand the risks and mortality of the species, and to improve the numbers returning to Canadian and American rivers where possible.

Another big current initiative is the Atlantic Salmon Trust's work on the Moray Firth Tracking Project in Scotland, part of the macro initiative, The Missing Salmon Project. Here, like on the Miramichi, they operate smolt traps including one on the Ballindalloch estate on the Avon. The traps, windmill-like, or rather waterwheel-like, turn in the flow and catch smolts in a box at the rear. From here they are taken out by scientific volunteers and helpers and in a very gently anaesthetised state, to prevent chance of damage, they are counted, measured and monitored. As part of the study, 150 smolts were fitted with acoustic monitors to try to determine mortality of smolts in the river, the idea also being that threats to their life are logged and tackled as part of the salmon conservation plan.

Finally, the Nith Salmon Smolt Tracking Project is monitoring Atlantic salmon smolts that live in the southwestern Scottish river and to find out what they experience during their migration to the coast. There are gaps in our knowledge about the smolts' journey, and for certain we know survival rates are impacted greatly and many of the fish may not make it as far as the river's estuary, and certainly not the open sea. Studies on other Scottish rivers have suggested that as many as 40-60 per cent of smolts leaving their tributaries never made it to sea. The journey is so especially dangerous because of the array of predators lining up to hunt the more visible silvered-up smolts.

The NDSFB puts it rather nicely: 'They have effectively gone from hiding in full camouflage to running a marathon in a high visibility onesie!'

> We believe that the risk of predation is made worse by man-made problems along the route such as the caul in Dumfries town centre that funnels the smolts into a choke point. You will most likely have seen for yourself the lines of birds that often wait on the caul for any fish unfortunate enough to swim too close! The predation risk may also be increased along stretches of the river that have been significantly altered. For example, areas where bankside trees and vegetation have been cleared or where the shape of the river and its banks have

been altered. Cover provided by overhanging plants and by naturally undercut banks is very important for the smolts and provides protection from aerial predators, removing this cover leaves them very exposed. By tracking the smolts through these stages and monitoring their loss rate we hope to discover which threats present the biggest risks, allowing us to prioritise areas for restoration work, hopefully easing the journey for future smolts.

The Miramichi Salmon Association in New Brunswick, Canada has conducted long-term smolt population estimates via a 'mark-recapture' scheme to assess the numbers of young salmon heading out to sea. This very practical method of tagging smolts then recapturing a proportion of the smolt run downstream 'has been shown to show' how many are tagged and gives a real time idea of smolt numbers.

How it works is that at upstream locations smolts are captured in smolt wheels. This is on the upper Miramichi and on tributaries like the Dungarvon. Then, further downstream, a cross section of the smolt population is recaptured and the proportion that are tagged indicates the overall smolt numbers, later extrapolated to numbers of fish that year.

> We have conducted smolt population estimates on the Miramichi River for many years using a mark-recapture system. A mark-recapture study works by applying tags to a number of smolts at an upstream location using the smolt wheels (rotary screw traps), and then by using a recapture platform further downstream, you can count how many tagged versus untagged smolts are captured at this location and produce an estimate of how many smolts are leaving that river system. The Miramichi River needs approximately 1.8 million smolts (1.2 million from the SW branch and 600,000 from the NW branch) migrating to the ocean annually to support a healthy population.

In 2014-2015, the smolt estimate study was expanded to encompass the entire Miramichi River (both branches) because of problems with washouts and freshets on the NW branch during the previous years' studies, which were giving inaccurate estimates. During these two years, four smolt wheels were installed on both branches of the Miramichi River: two on the Northwest (Trout Brook, and the mouth of the Sevogle) and two on the Southwest (Rocky Brook and the Cains). A new, larger trap net was constructed and installed near the Centennial Bridge on the Chatham side of the River as our recapture platform. The results from these studies are available below.

Meanwhile, the Nith Project has been busy.

> This year the Nith Catchment Fishery Trust will be tagging 50 of the salmon smolts leaving two of our main tributaries, the Mennock and Crawick waters as

part of a project called the Nith Salmon Smolt Tracking Project. We are doing so in the hope of increasing our knowledge as to how quickly the smolts migrate down river and which points along the route the smolts are most at risk, with the hope of eventually being able to implement measures to ease their passing in any identified danger zones. The Nith Salmon Smolt Tracking Project (NSSTP) project is supported by the Holywood Trust, Dumfries and Galloway Council and Nith District Salmon Fishery Board. During this spring we will track the salmon smolts through 8 different relay points along the river until they reach the Solway Firth near New Abbey, but that's not all! When the smolts leave the river and head out to sea our project will merge with a larger nationwide project called the West Coast Salmon Smolt Tracking Project (WCSSTP). The WCSSTP is designed to again increase our knowledge about the smolts journey, this time about the routes they take and the dangers they face while travelling round the west coast of Scotland. Seven other Scottish fisheries trusts will be taking part in the WCSSTP which is being run in co-operation with the Scottish Government and the Atlantic Salmon Trust. For the WCSSTP we will tag a further 100 smolts as they enter the lower reaches of the river. We hope that the results of our project and those of the WCSSTP will help us better understand the Atlantic salmon's migration and possibly allow us to reduce smolt mortality rates, hopefully resulting in more adults eventually returning home to spawn, reducing some of the pressure on the currently endangered species. Over the next few months we will be posting a lot more updates and details about the project, so keep your eyes peeled!

So this disparate effort – is it wrong? As always, it is good for science to work together on a global, macro scale. But in the case of *Salmo salar*, fragmentation to individual rivers has its benefit.

APPENDIX FOUR

The Eternal Salmon

Like most creatures, salmon feature widely in folk lore and in many cases the stories are pretty wild, with a salmon giving a ride to a dozen people on its back, persuading husbands to be good to their wives, stop beating their children and other stories. All of these, at the time, were to make a point and often the salmon is chosen almost at random – the same tales can be found featuring deer, boar, even ants. But this tale from the Yakima people who were first identified as a separate tribe in the northern United States in 1750 (although they had been there for millennia before, of course) is a warning to them and ultimately to everyone, that natural resources cannot be taken for granted, no matter how persuasive the argument. 'The People' is how most tribes described themselves, it is not unique to the Yakima. With no written language, the tale is simple, so it can be passed on orally. There are few frills and additions, because they would be lost in the telling and the message might become blurred. So the story below is in the style of, but not the original words of, the Yakima.

The Creator instructed The People in how to care for his creation. They could live in harmony with Creation by remembering and following what He told them. The Creator told The People that Salmon had been created especially for them. He told them to take special care to always show respect for them and for their well-being. The Creator told The People, 'Be sure that you only catch as much Salmon as you need to eat, and you will always have plenty.'

For a long while, this went well, with The People living according to these good words, and in harmony with their surroundings and the Salmon. There was always plenty of food. However, there came a time when The People forgot their role in living in harmony. The Salmon stocks had never let them down, there had been no years of want and so they assumed that the fish would always be there for them, and they thought that harmony did not require anything from them. The People sometimes wasted the Salmon, and then neglected to dry all they had caught, so that they needed to catch more and more, just to have what they thought was enough. The young who grew up this way strayed further from the Creator's good words and further from harmony. The Coyote spoke to the youngsters and his words led them astray. It wasn't his way to protect and conserve and as far as they could tell, he seemed to be doing all right. While they lived like this, time passed quickly, and suddenly one day, or so it seemed, they found there were no Salmon! The People walked up and down the

Gaula salmon. (HG)

Hedge End, an upland Nith salmon pool, in October 2020. (HG)

river, looking for Salmon. Everyone was perpetually hungry, and it was a sad time with few answers.

One day, they came upon a dead Salmon on the bank of the river. They remembered the good words of The Creator and realized what they had done. They called a council to talk together about how they could correct the damage, if that was even possible. They talked about how in past times, or so they had heard, those with supernatural powers could bring life back by stepping over a creature five times. Each one of The People tried stepping over the salmon five times, and each time it was the same. The Salmon was as dead as ever.

Then someone remembered that nobody had asked Old Man Rattlesnake to step over the salmon five times. Old Man Rattlesnake lived by himself and stayed alone so they sent a runner to fetch the old man and tell him that The People needed his help to bring back the Salmon. The old man asked the runner, 'What makes you think I have supernatural powers?' The runner had expected this, so said, 'You are our only hope. Everyone else has tried and failed.' It took a bit of doing, but finally the old man was persuaded to come, but because he was so old, it took him a long time to make his way to The People.

World salmon, showing us all the way. (Illustration by hiromikitakibi)

While they were waiting, Coyote tried his best to convince The People that *he* had supernatural powers. He told them to watch while he stepped over the Salmon. and on the fifth time stepping over the Salmon, he moved it with his foot. It wasn't the result they were looking for, in that the Salmon didn't flop back into the water and swim away or anything; it was just a twitch. But still Coyote tried to convince the people. 'Look! Did you see the salmon move? I am indeed supernatural and I have caused Salmon to return!' But The People had finally realized that Coyote was tricking them and that following him had led them to neglect The Creator's advice so they didn't listen when Coyote kept on telling them he had revived Salmon.

Finally, Old Man Rattlesnake arrived. Slowly, he made his way to salmon, and stepped solemnly over the body. The People held their breath but, on the fifth time, the supernatural powers they had hoped for showed themselves; Old Man Rattlesnake suddenly disappeared and went into Salmon and once again the rivers were full of fish.

From that time, The People committed themselves to living in harmony with Salmon. They remembered to follow the good advice of The Creator, to take only what they needed, to carefully husband any extra they didn't need at once and to live in harmony with the world. And they knew The Creator had told them to show their children how to live this way, remembering the example of their grandmothers and grandfathers.

Bibliography

Buller, Fred and Falkus, Hugh, *Freshwater Fishing*, Grange Books, 1994
Buller, Fred, *Fred Buller's Domesday Book of Giant Salmon*, Constable, 2007
Burns, Brad, *Closing the Season*, Burns Flyfishing, 2013
Burns, Brad, *On the Cains*, Stackpole Books, 2020
Coates, Peter, *Salmon*, Reaktion Books, 2006
Greenhalgh, Malcolm, *The Complete Salmon Fisher, volume 1*, Blandford, 1996
Mills, Derek, *Salmon and Science*, Coch-y-Bonddu Books Ltd, 2016
Pearce, Fred, *A Trillion Trees*, Grants Books, 2021
Wigan, Michael, *The Salmon*, William Collins, 2013

Index

A
Alaska 21, 59
Alevins 36
Anthropocene 62
Atlantic salmon 24, 50, 53, 62, 92, 93, 96, 97, 98, 99, 103, 111
Atlantic Salmon Trust 133
Aquaculture 51-53
Aquaculture escapes 54
Arctic 64, 68

B
British Columbia 60
Buckland, Frank 17

C
Cains river NB 106
Canada (western) 16, 24, 86, 95, 113;
Canada (eastern) 22, 24, 106, 120
Capelin 44
Chinook salmon 25, 59, 96, 97, 98, 100
Chum salmon 25, 96, 98, 100
Climate and climate change 33, 62, 64, 116
Coho salmon 101
COP26 62

E
Eggs (salmon) 28, 46
El Nino 24, 71
European Union 77, 78

F
Faroe islands 43, 44
Faroese science 35
First Nation communities 16, 27
Fish-eating birds 42
Folk memory 18
Forestry (and deforestation and logging) 87, 88, 90
Freshwater Fisheries Laboratory, Pitlochry 39
Fry 28

G
Gaula river 26, 115
Gray, Peter (Tyne hatchery manager) 57-60, 77
Greenland 32, 43, 44
Grilse 29

I
Industrialisation 18

K
Katmai national park, Alaska 36
Kelts 49, 132
Kola Peninsula 96, 104

M
Maclean, Norman 111
Maine salmon rivers 77, 86, 97

Masu or cherry salmon 102
Mining impact on salmon rivers 71
Miramichi Salmon Association 55-56, 104, 134
Missing Salmon Alliance (Likely Suspects Framework) 42, 109, 114
Mixed-stock fishery 21
Moray firth 32

N
NASCO 111
NASF 37
Nith (fish stocks, hatchery) 56-57, 74, 104, 133
North America 16
Norse mythology 18
Nova Scotia 33
NPAFC 111

O
Obstacles and dams 83-84
Ocean temperatures 39
Oregon 18
Orkla 29
Osenka 39

P
Pacific ocean 25
Parr 28, 42, 49
Pink salmon 25, 96, 98, 99, 101, 102
Poaching 90, 91
Post-smolts 29

R
Ranching 55
Redds 61

River temperature/salmon behaviour 45
Roosevelt, President Theodore 18-20

S
Sea levels 71
Sea-trout 44
Seventies pollution 77
Skokomish river 25
Smolts (smolt traps) 28, 74, 109 (Feshie), 132-135 (smolt trackers)
Sockeye salmon 24, 59, 96, 99, 101
Spawning 46, 61
Sturlaugsson, Johannes 33-34
Sub-arctic 32, 68

T
Taiwanese salmon 102
Thames salmon 78
Thunberg, Greta 67-68, 70
Thurso fish shape 55
Trondheim fjord 29

V
Varzuga river 110
Vigfusson, Orri 37, 115

W
Wildfires 87
WWF 87
Wyville-Thomson Ridge 35

Z
Zooplankton 33